BIG IDEAS MATH®
Algebra 2

Student Journal

- Maintaining Mathematical Proficiency
- Exploration Journal
- Notetaking with Vocabulary
- Extra Practice

Erie, Pennsylvania

Photo Credits

Cover Image CVADRAT/Shutterstock.com

283 *top right* Sascha Burkard/Shutterstock.com;
center Ruslan Gi/Shutterstock.com;
284 Sanchai Khudpin/Shutterstock.com,
cobalt88/Shutterstock.com;
308 Sascha Burkard/Shutterstock.com;
327 Denis Cristo/Shutterstock.com;
330 sauletas/Shutterstock.com

ISBN 13: 978-1-60840-854-2
ISBN 10: 1-60840-854-X

56789-VLP-18 17 16 15

Contents

Contents

Contents

Contents

Contents

Contents

Contents

Chapter 11 Data Analysis and Statistics

About the Student Journal

Maintaining Mathematical Proficiency

The Maintaining Mathematical Proficiency corresponds to the Pupil Edition Chapter Opener. Here you have the opportunity to practice prior skills necessary to move forward.

Exploration Journal

The Exploration pages correspond to the Explorations and accompanying exercises in the Pupil Edition. Here you have room to show your work and record your answers.

Notetaking with Vocabulary

This student-friendly notetaking component is designed to be a reference for key vocabulary, properties, and core concepts from the lesson. There is room to add definitions in your words and take notes about the core concepts.

Extra Practice

Each section of the Pupil Edition has an additional Practice with room for you to show your work and record your answers.

Name______________________________ Date__________

Chapter 1 Maintaining Mathematical Proficiency

Evaluate.

1. $7 \bullet 3^2 + 11$

2. $10 - 3(3 + 1)^3$

3. $64 \div 4^2 + \frac{1}{2}$

4. $-99 \div 3^2 \bullet 5$

5. $\frac{1}{7}\left(7^2 + 28\right)$

6. $-\frac{1}{8}(8 + 24) - 2^2$

Graph the transformation of the figure.

7. Translate the rectangle 3 units left and 4 units up.

8. Reflect the right triangle in the y-axis. Then translate 3 units down.

9. Translate the trapezoid 2 units up. Then reflect in the x-axis.

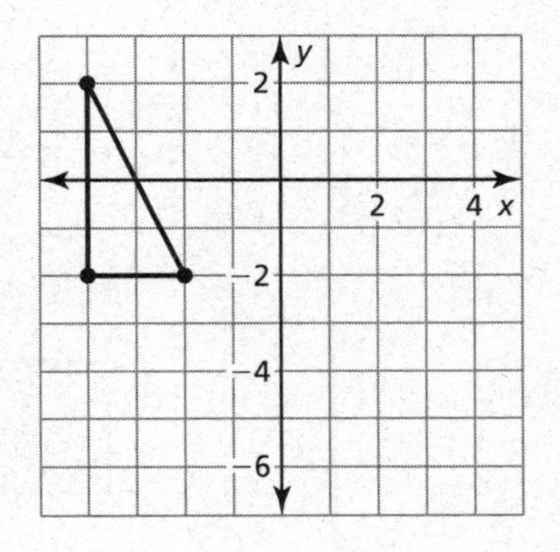

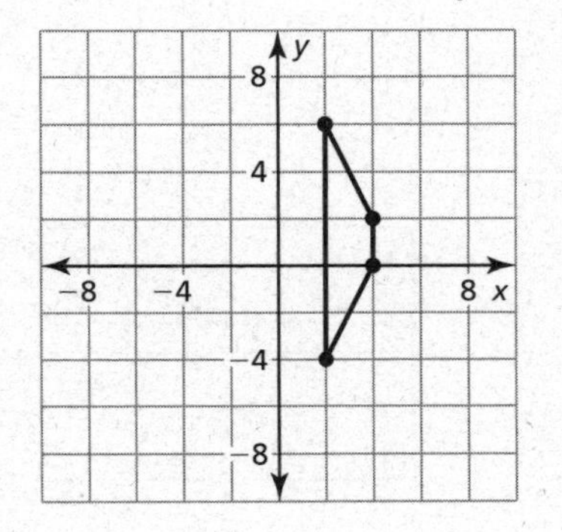

10. The point $(1, 1)$ is on $f(x)$. After a series of 3 transformations, $(1, 1)$ has been moved to $(2, -7)$. Write a function $g(x)$ that represents the transformations on $f(x)$.

Name ______________________________ Date __________

1.1 Parent Functions and Transformations

For use with Exploration 1.1

Essential Question What are the characteristics of some of the basic parent functions?

EXPLORATION: Identifying Basic Parent Functions

Work with a partner. Graphs of eight basic parent functions are shown below. Classify each function as *constant*, *linear*, *absolute value*, *quadratic*, *square root*, *cubic*, *reciprocal*, or *exponential*. Justify your reasoning.

a. **b.**

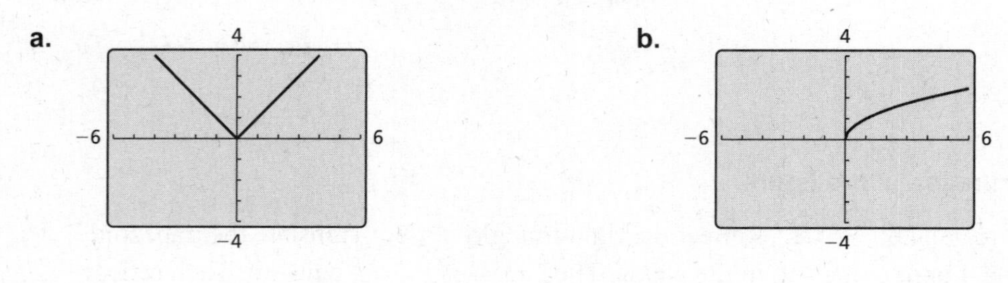

c.

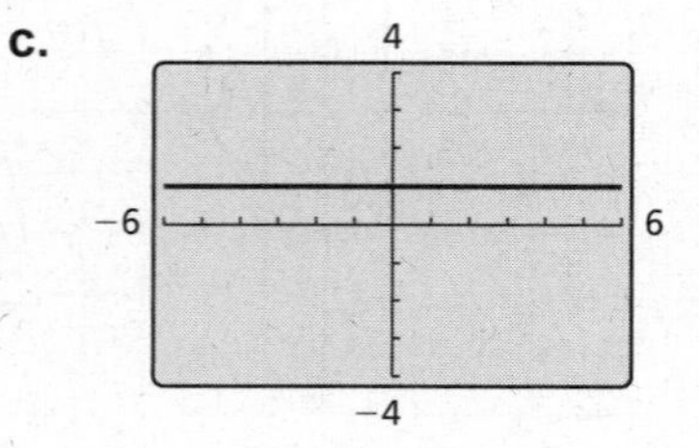

d.

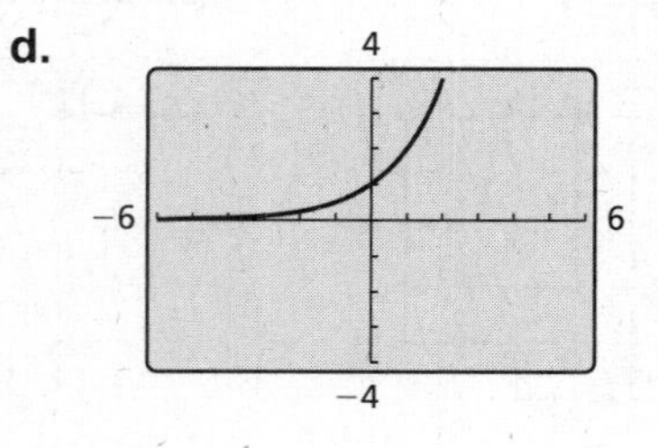

e.

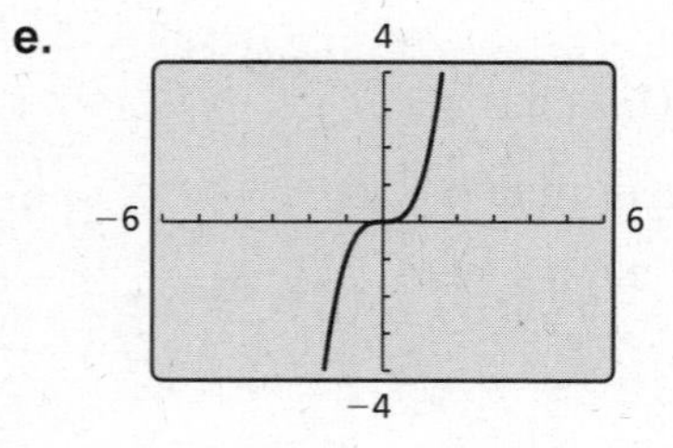

f.

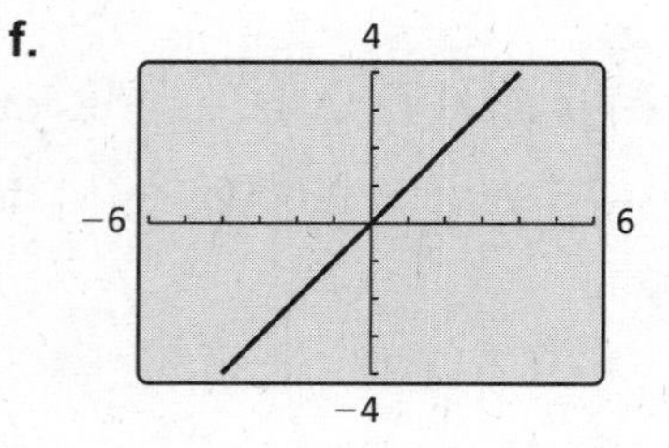

Name____ Date__________

1.1 Parent Functions and Transformations (continued)

1 EXPLORATION: Identifying Basic Parent Functions (continued)

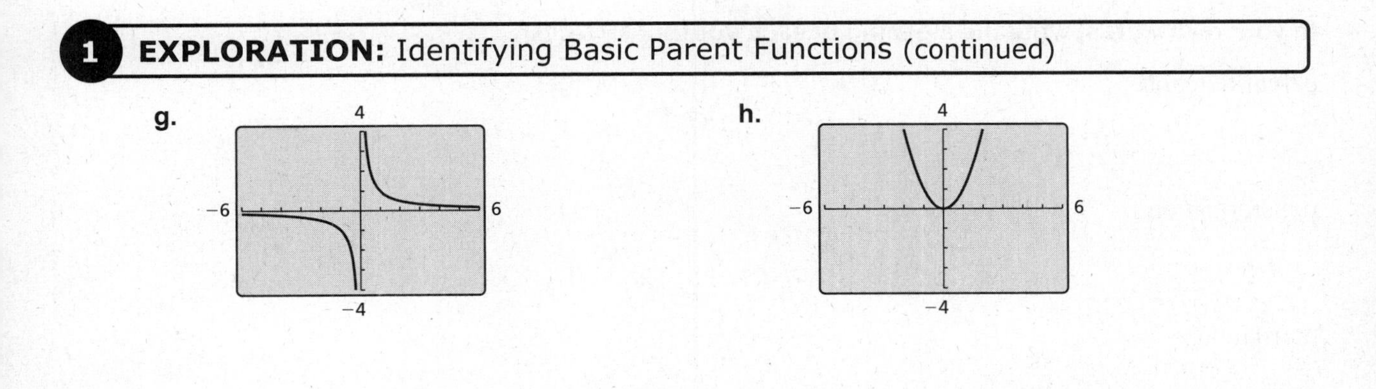

Communicate Your Answer

2. What are the characteristics of some of the basic parent functions?

3. Write an equation for each function whose graph is shown in Exploration 1. Then use a graphing calculator to verify that your equations are correct.

Name ______________________________ Date __________

1.1 Notetaking with Vocabulary
For use after Lesson 1.1

In your own words, write the meaning of each vocabulary term.

parent function

transformation

translation

reflection

vertical stretch

vertical shrink

Core Concepts

Parent Functions

Family	Constant	Linear	Absolute Value	Quadratic
Rule	$f(x) = 1$	$f(x) = x$ 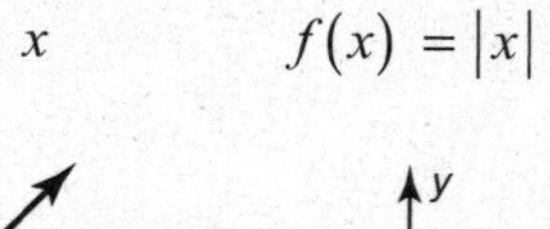	$f(x) = \lvert x \rvert$	$f(x) = x^2$
Graph	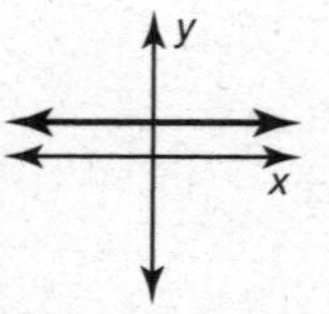	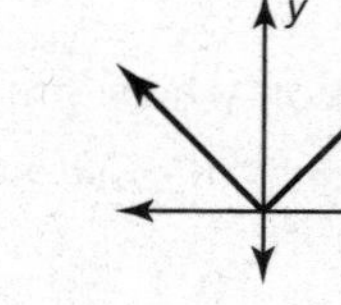		
Domain	All real numbers	All real numbers	All real numbers	All real numbers
Range	$y = 1$	All real numbers	$y \geq 0$	$y \geq 0$

Notes:

Name ______________________ Date __________

1.1 Notetaking with Vocabulary (continued)

Extra Practice

In Exercises 1–4, identify the function family to which *f* belongs. Compare the graph of *f* to the graph of its parent function.

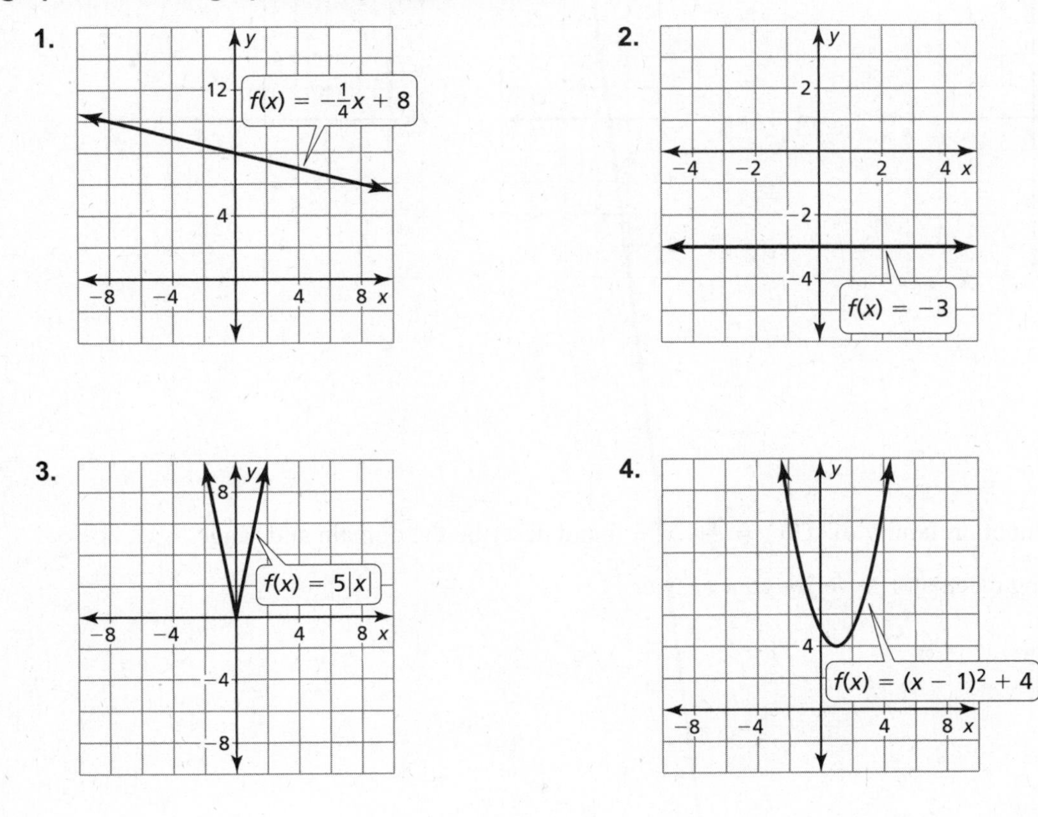

In Exercises 5–10, graph the function and its parent function. Then describe the transformation.

5. $f(x) = x - 7$

6. $f(x) = -9$

7. $f(x) = |x| + 1$

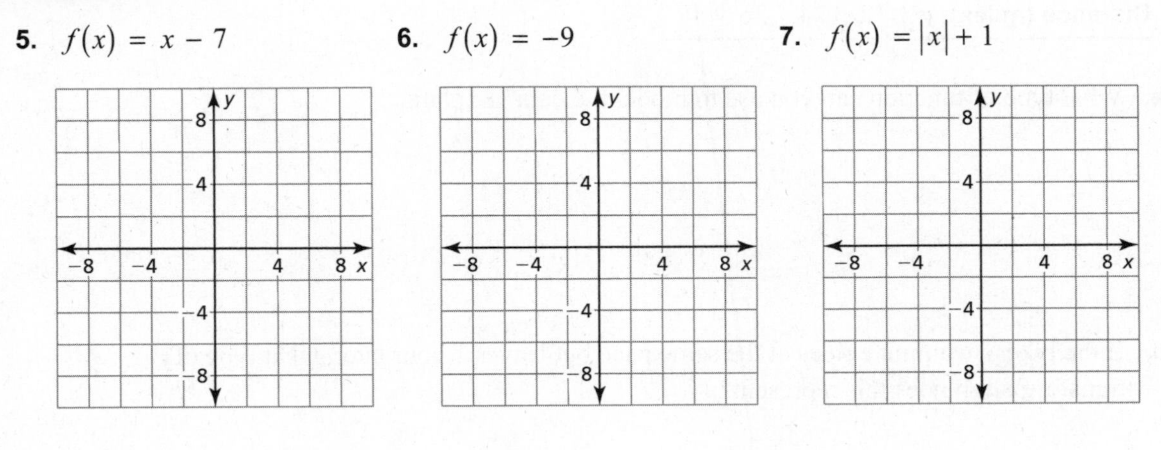

8. $h(x) = -x^2$

9. $f(x) = \frac{1}{8}x^2$

10. $g(x) = 6|x|$

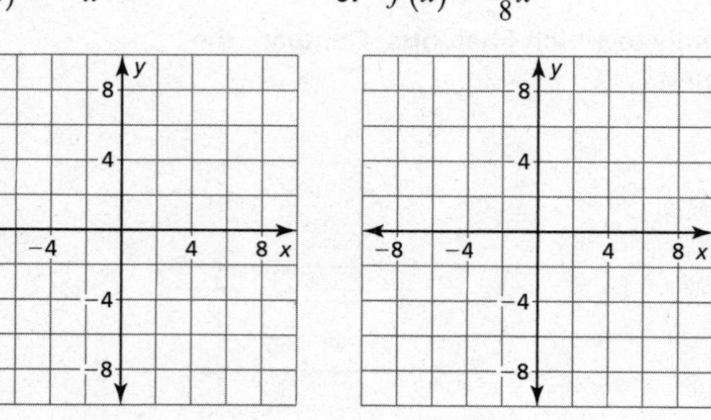

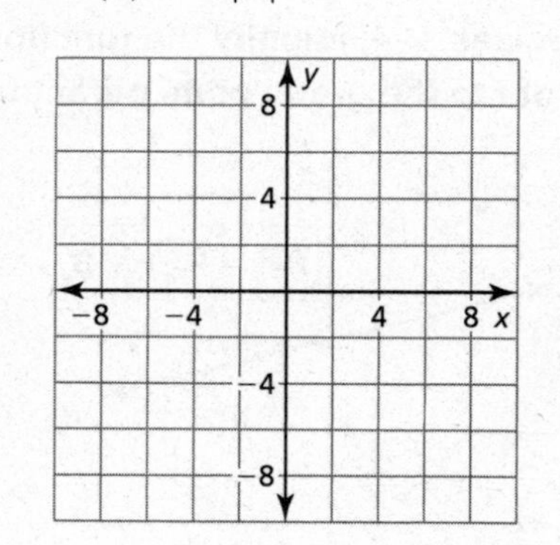

11. Identify the function family of $f(x) = \frac{1}{3}|-x| + 4$ and describe the domain and range. Use a graphing calculator to verify your answer.

12. The table shows the distance a biker rides in his first team relay competition.

Time (hours), *x*	1	2	3	4
Distance (miles), *y*	12	24	36	48

a. What type of function can you use to model the data? Explain.

b. If the biker's teammate rides at the same pace but leaves 1 hour later, what type of transformation does this represent?

Name____ Date__________

1.2 Transformations of Linear and Absolute Value Functions

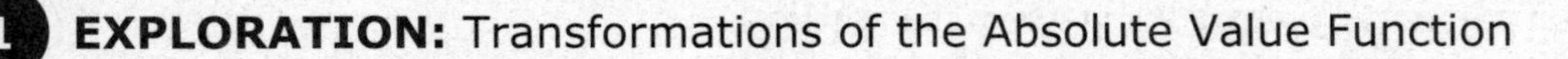

For use with Exploration 1.2

Essential Question How do the graphs of $y = f(x) + k$, $y = f(x - h)$, and $y = -f(x)$ compare to the graph of the parent function f?

1 EXPLORATION: Transformations of the Absolute Value Function

Go to *BigIdeasMath.com* for an interactive tool to investigate this exploration.

Work with a partner. Compare the graph of the function

$y = |x| + k$ Transformation

to the graph of the parent function

$f(x) = |x|.$ Parent function

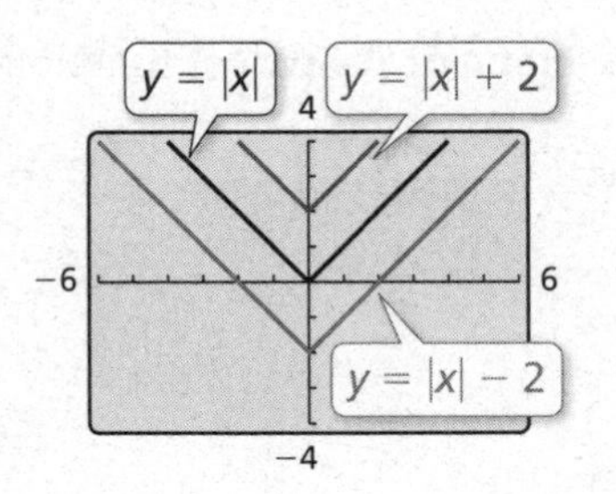

2 EXPLORATION: Transformations of the Absolute Value Function

Go to *BigIdeasMath.com* for an interactive tool to investigate this exploration.

Work with a partner. Compare the graph of the function

$y = |x - h|$ Transformation

to the graph of the parent function

$f(x) = |x|.$ Parent function

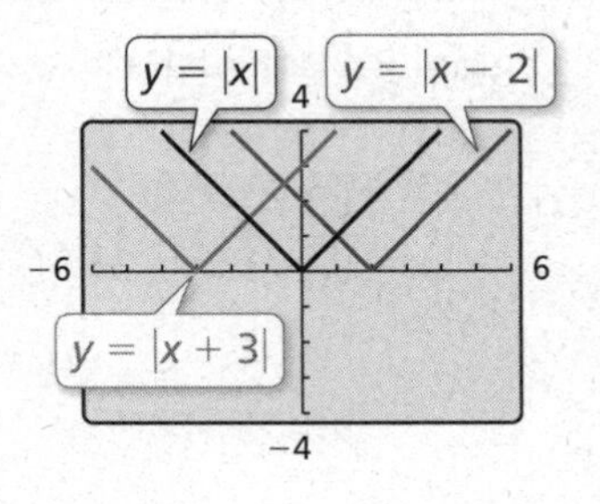

3 EXPLORATION: Transformation of the Absolute Value Function

Go to *BigIdeasMath.com* for an interactive tool to investigate this exploration.

Work with a partner. Compare the graph of the function

$y = -|x|$ Transformation

to the graph of the parent function

$f(x) = |x|.$ Parent function

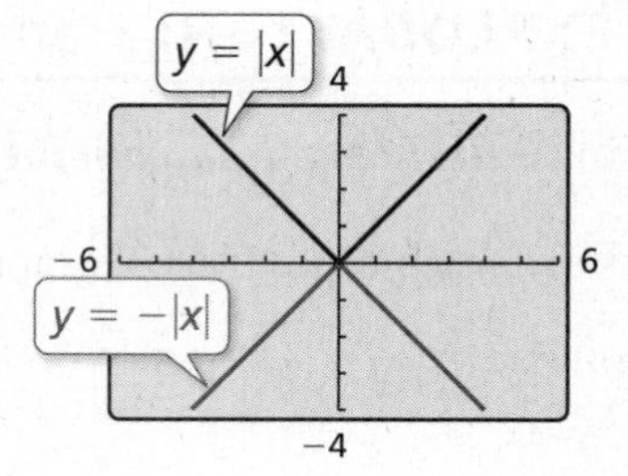

Communicate Your Answer

4. How do the graphs of $y = f(x) + k$, $y = f(x - h)$, and $y = -f(x)$ compare to the graph of the parent function f?

5. Compare the graph of each function to the graph of its parent function f. Use a graphing calculator to verify your answers are correct.

a. $y = \sqrt{x} - 4$ **b.** $y = \sqrt{x + 4}$ **c.** $y = -\sqrt{x}$

d. $y = x^2 + 1$ **e.** $y = (x - 1)^2$ **f.** $y = -x^2$

Name______________________________ Date__________

1.2 Notetaking with Vocabulary

For use after Lesson 1.2

Core Concepts

Horizontal Translations

The graph of $y = f(x - h)$ is a horizontal translation of the graph of $y = f(x)$, where $h \neq 0$.

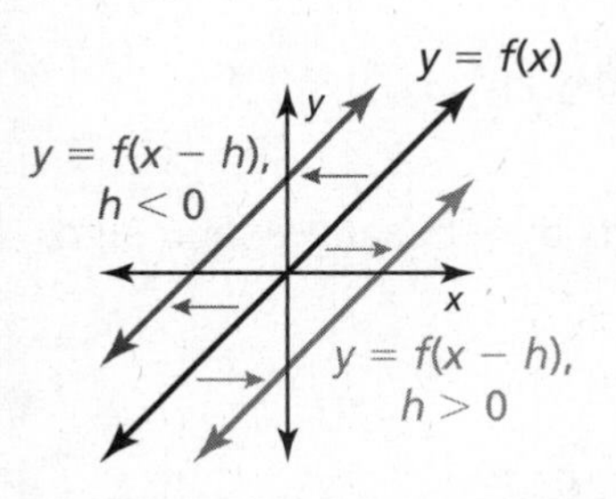

Subtracting h from the **inputs** before evaluating the function shifts the graph left when $h < 0$ and right when $h > 0$.

Vertical Translations

The graph of $y = f(x) + k$ is a vertical translation of the graph of $y = f(x)$, where $k \neq 0$.

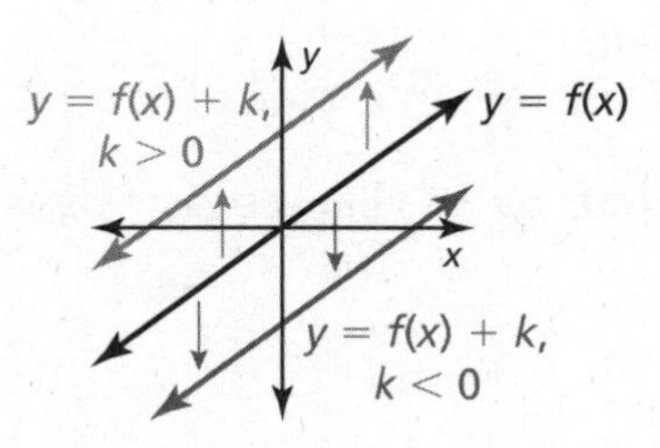

Adding k to the **outputs** shifts the graph down when $k < 0$ and up when $k > 0$.

Notes:

1.2 Notetaking with Vocabulary (continued)

Reflections in the *x*-axis

The graph of $y = -f(x)$ is a reflection in the x-axis of the graph of $y = f(x)$.

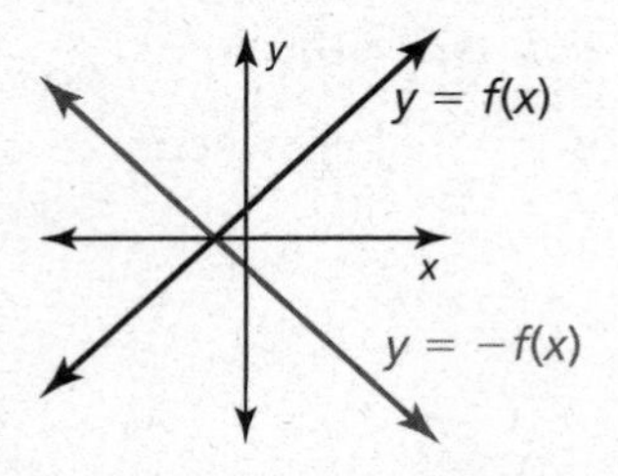

Multiplying the **outputs** by -1 changes their signs.

Reflections in the *y*-axis

The graph of $y = f(-x)$ is a reflection in the y-axis of the graph of $y = f(x)$.

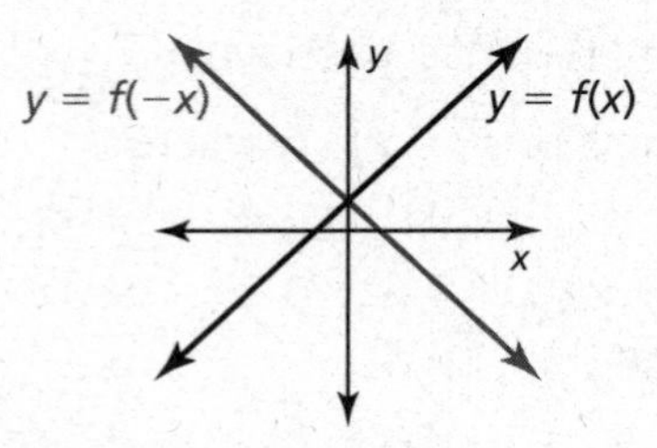

Multiplying the **inputs** by -1 changes their signs.

Notes:

Horizontal Stretches and Shrinks

The graph of $y = f(ax)$ is a horizontal stretch or shrink by a factor of $\frac{1}{a}$ of the graph of $y = f(x)$, where $a > 0$ and $a \neq 1$.

Multiplying the **inputs** by a before evaluating the function stretches the graph horizontally (away from the y-axis) when $0 < a < 1$, and shrinks the graph horizontally (toward the y-axis) when $a > 1$.

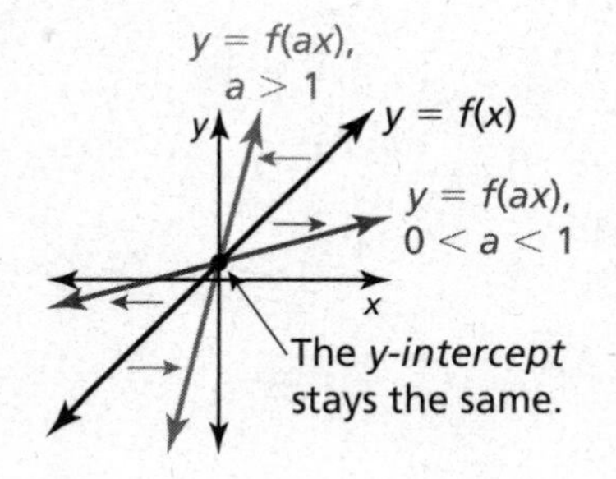

Notes:

1.2 Notetaking with Vocabulary (continued)

Vertical Stretches and Shrinks

The graph of $y = a \bullet f(x)$ is a vertical stretch or shrink by a factor of a of the graph of $y = f(x)$, where $a > 0$ and $a \neq 1$.

Multiplying the **outputs** by a stretches the graph vertically (away from the x-axis) when $a > 1$, and shrinks the graph vertically (toward the x-axis) when $0 < a < 1$.

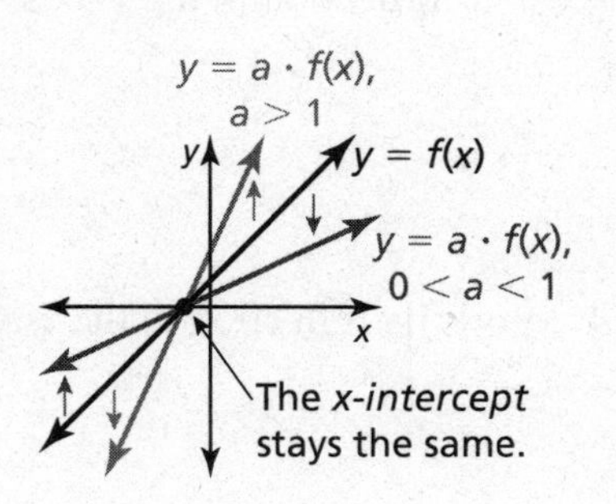

Notes:

Extra Practice

In Exercises 1–9, write a function *g* whose graph represents the indicated transformation of the graph of *f*. Use a graphing calculator to check your answer.

1. $f(x) = \left|\frac{1}{3}x\right|$; translation 2 units to the left

2. $f(x) = -|x + 9| - 1$; translation 6 units down

3. $f(x) = -2x + 2$; translation 7 units down

Name ______________________ Date __________

1.2 Notetaking with Vocabulary (continued)

4. $f(x) = \frac{1}{2}x + 8$; reflection in the x-axis

5. $f(x) = 4 + |x + 1|$; reflection in the y-axis

6. $f(x) = -5x$; vertical shrink by a factor of $\frac{1}{5}$

7. $f(x) = |x + 3| + 2$; vertical stretch by a factor of 4

8. $f(x) = 3x - 9$; horizontal stretch by a factor of 6

9. $f(x) = -|8x| - 4$; horizontal shrink by a factor of $\frac{1}{4}$

10. Consider the function $f(x) = |x|$. Write a function g whose graph represents a reflection in the x-axis followed by a horizontal stretch by a factor of 3 and a translation 5 units down of the graph of f.

11. Which of the transformation(s) in Section 1.2 will *not* change the y-intercept of $f(x) = |x| + 3$?

Name____ Date__________

1.3 Modeling with Linear Functions

For use with Exploration 1.3

Essential Question How can you use a linear function to model and analyze a real-life situation?

1 EXPLORATION: Modeling with a Linear Function

Go to *BigIdeasMath.com* for an interactive tool to investigate this exploration.

Work with a partner. A company purchases a copier for $12,000. The spreadsheet shows how the copier depreciates over an 8-year period.

	A	B
1	Year, t	Value, V
2	0	$12,000
3	1	$10,750
4	2	$9,500
5	3	$8,250
6	4	$7,000
7	5	$5,750
8	6	$4,500
9	7	$3,250
10	8	$2,000
11		

a. Write a linear function to represent the value V of the copier as a function of the number t of years.

b. Sketch a graph of the function. Explain why this type of depreciation is called *straight line depreciation.*

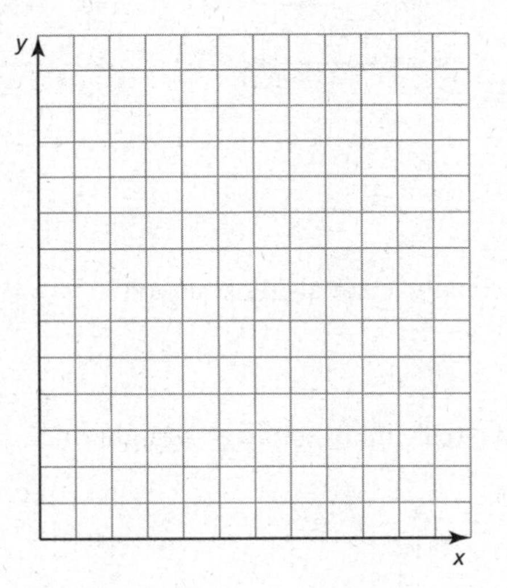

c. Interpret the slope of the graph in the context of the problem.

Name ____ Date __________

2 EXPLORATION: Modeling with Linear Functions

Work with a partner. Match each description of the situation with its corresponding graph. Explain your reasoning.

a. A person gives \$20 per week to a friend to repay a \$200 loan.

b. An employee receives \$12.50 per hour plus \$2 for each unit produced per hour.

c. A sales representative receives \$30 per day for food plus \$0.565 for each mile driven.

d. A computer that was purchased for \$750 depreciates \$100 per year.

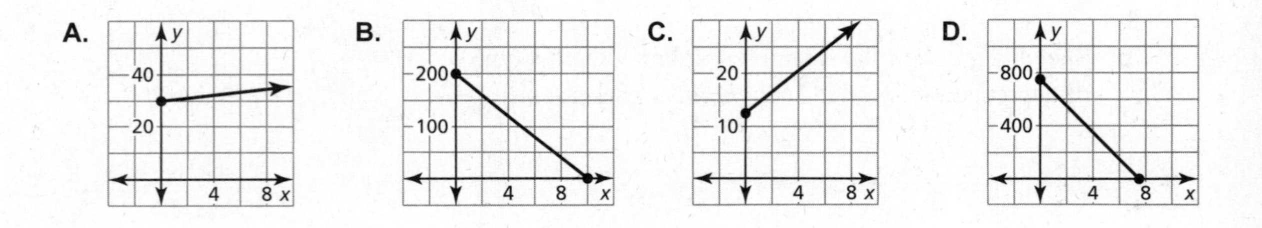

Communicate Your Answer

3. How can you use a linear function to model and analyze a real-life situation?

4. Use the Internet or some other reference to find a real-life example of straight line depreciation.

a. Use a spreadsheet to show the depreciation.

b. Write a function that models the depreciation.

c. Sketch a graph of the function.

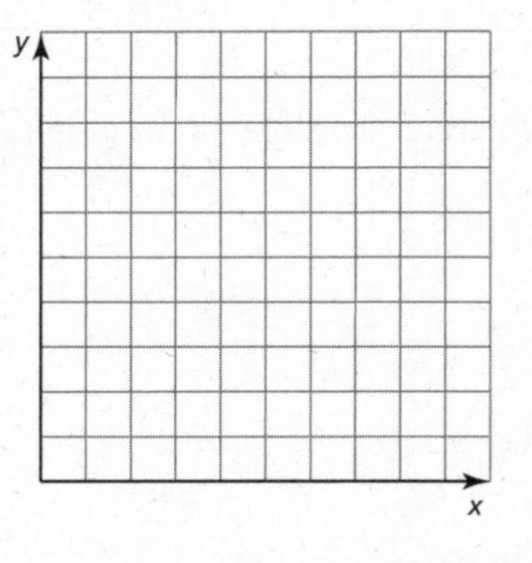

Name____ Date__________

1.3 Notetaking with Vocabulary

For use after Lesson 1.3

In your own words, write the meaning of each vocabulary term.

line of fit

line of best fit

correlation coefficient

Core Concepts

Writing an Equation of a Line

Given slope m and y-intercept b Use slope-intercept form:

$$y = mx + b$$

Given slope m and a point (x_1, y_1) Use point-slope form:

$$y - y_1 = m(x - x_1)$$

Given points (x_1, y_1) and (x_2, y_2) First use the slope formula to find m. Then use point-slope form with either given point.

Notes:

1.3 Notetaking with Vocabulary (continued)

Finding a Line of Fit

Step 1 Create a scatter plot of the data.

Step 2 Sketch the line that most closely appears to follow the trend given by the data points. There should be about as many points above the line as below it.

Step 3 Choose two points on the line and estimate the coordinates of each point. These points do not have to be original data points.

Step 4 Write an equation of the line that passes through the two points from Step 3. This equation is a model for the data.

Notes:

Extra Practice

In Exercises 1–3, use the graph to write an equation of the line and interpret the slope.

1. **2.** **3.**

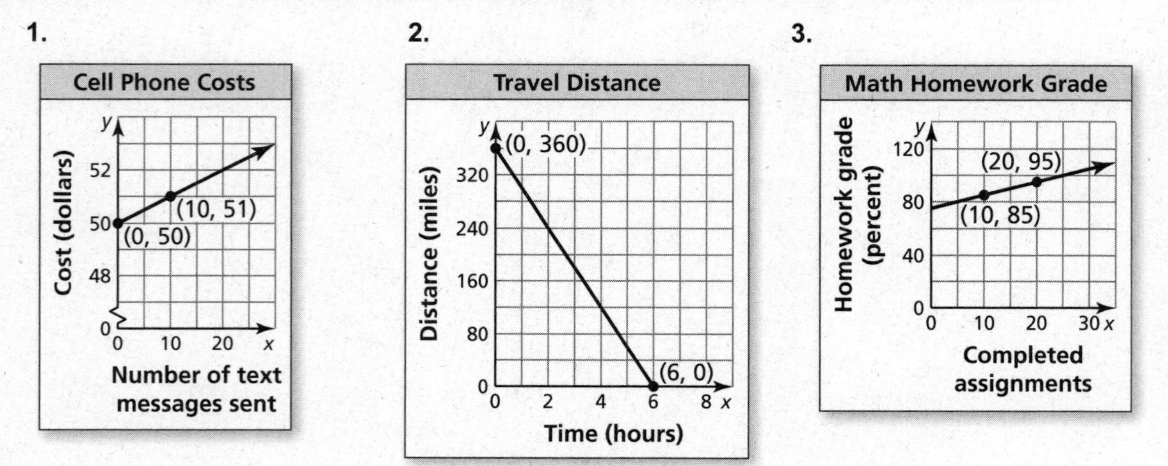

Name____ Date__________

1.3 Notetaking with Vocabulary (continued)

4. The cost of parking in a parking garage in Chicago is represented by the equation $y = 15x + 20$ where y is the total cost (in dollars) and x is the time (in hours). The table shows the total cost to park in a parking garage in Denver. Which city's parking garage charges more per hour and by how much more? After how many hours would parking in both cities cost the same?

Hours, *x*	2	3	4	5
Cost, *y*	43	51	59	67

In Exercises 5–7, use the *linear regression* feature on a graphing calculator to find an equation of the line of best fit for the data. Find and interpret the correlation coefficient.

5. **6.** **7.**

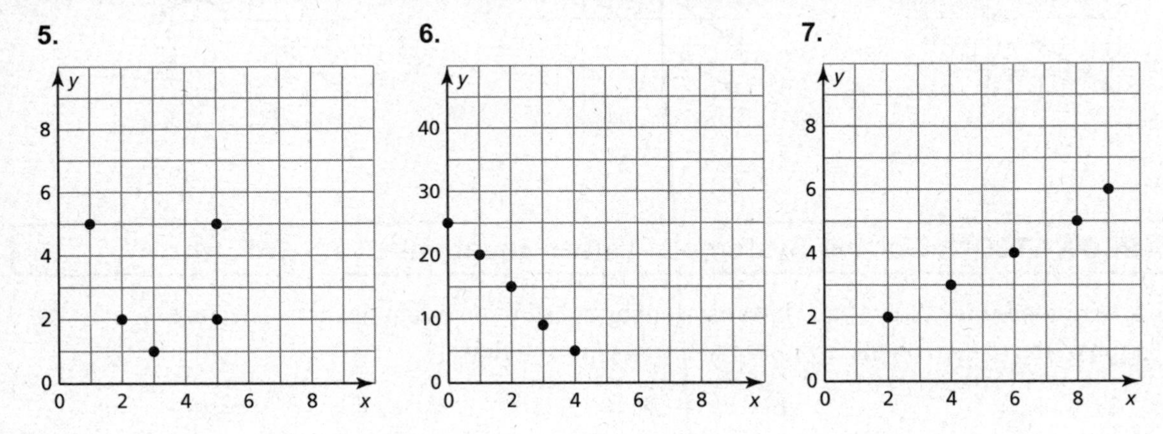

Name ______________________________ Date __________

1.4 Solving Linear Systems

For use with Exploration 1.4

Essential Question How can you determine the number of solutions of a linear system?

A linear system is *consistent* when it has at least one solution. A linear system is *inconsistent* when it has no solution.

1 EXPLORATION: Recognizing Graphs of Linear Systems

Work with a partner. Match each linear system with its corresponding graph. Explain your reasoning. Then classify the system as *consistent* or *inconsistent*.

a. $2x - 3y = 3$
$-4x + 6y = 6$

b. $2x - 3y = 3$
$x + 2y = 5$

c. $2x - 3y = 3$
$-4x + 6y = -6$

A.

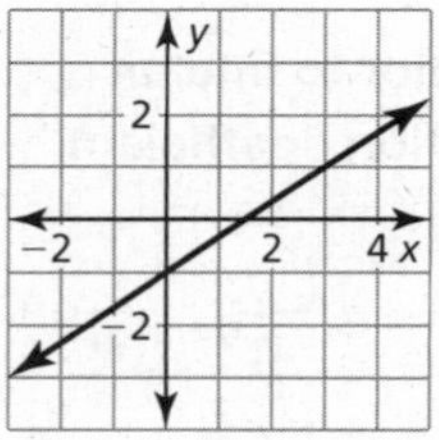

B.

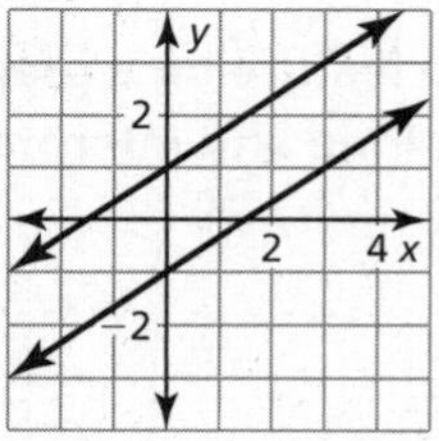

C.

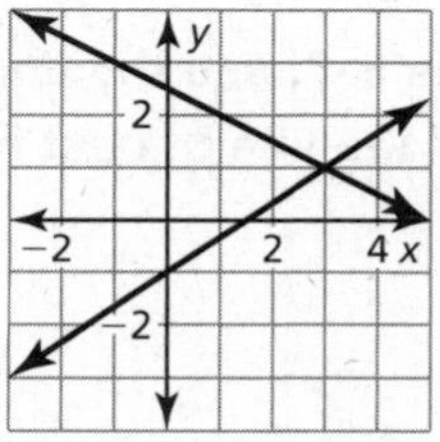

2 EXPLORATION: Solving Systems of Linear Equations

Work with a partner. Solve each linear system by substitution or elimination. Then use the graph of the system on the next page to check your solution.

a. $2x + y = 5$
$x - y = 1$

b. $x + 3y = 1$
$-x + 2y = 4$

c. $x + y = 0$
$3x + 2y = 1$

Name______________________________ Date__________

1.4 Solving Linear Systems (continued)

2 EXPLORATION: Solving Systems of Linear Equations (continued)

a.

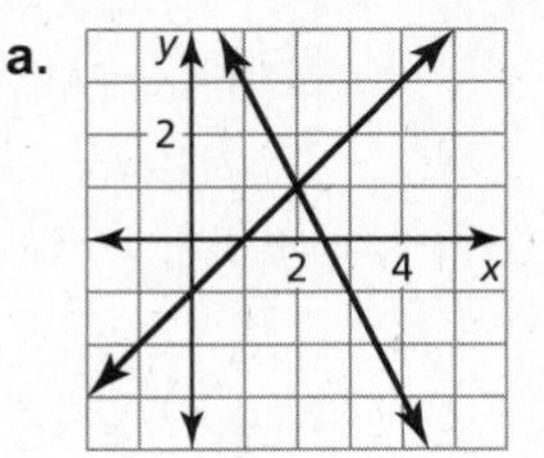

b.

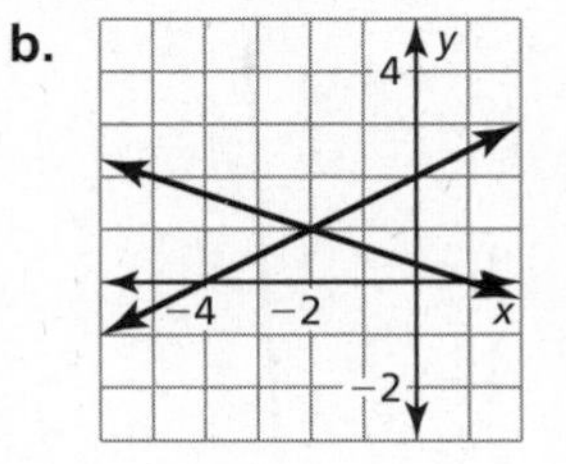

c.

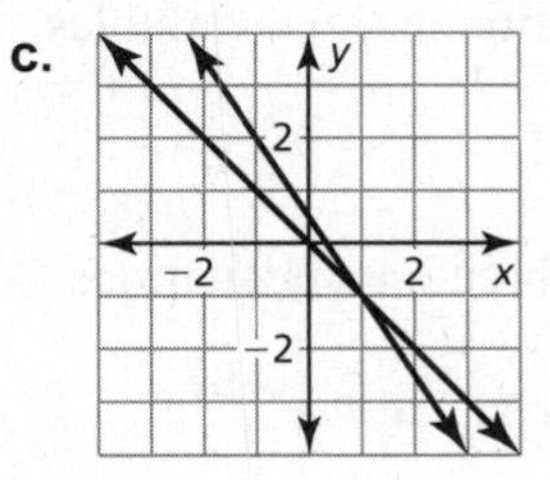

Communicate Your Answer

3. How can you determine the number of solutions of a linear system?

4. Suppose you were given a system of *three* linear equations in *three* variables. Explain how you would approach solving such a system.

5. Apply your strategy in Question 4 to solve the linear system.

$$\begin{aligned} x + y + z &= 1 && \text{Equation 1} \\ x - y - z &= 3 && \text{Equation 2} \\ -x - y + z &= -1 && \text{Equation 3} \end{aligned}$$

Name ______________________________ Date __________

Notetaking with Vocabulary

For use after Lesson 1.4

In your own words, write the meaning of each vocabulary term.

linear equation in three variables

system of three linear equations

solution of a system of three linear equations

ordered triple

Core Concepts

Solving a Three-Variable System

Step 1 Rewrite the linear system in three variables as a linear system in two variables by using the substitution or elimination method.

Step 2 Solve the new linear system for both of its variables.

Step 3 Substitute the values found in Step 2 into one of the original equations and solve for the remaining variable.

When you obtain a false equation, such as $0 = 1$, in any of the steps, the system has no solution.

When you do not obtain a false equation, but obtain an identity such as $0 = 0$, the system has infinitely many solutions.

Notes:

Name____ Date__________

1.4 Notetaking with Vocabulary (continued)

Extra Practice

In Exercises 1–3, solve the system using the elimination method.

1.
$$\begin{aligned} x + 2y - 3z &= 11 \\ 2x + y - 2z &= 9 \\ 4x + 3y + z &= 16 \end{aligned}$$

2.
$$\begin{aligned} x - y + 3z &= 19 \\ -2x + 2y - 6z &= 9 \\ 3x + 5y + 2z &= 3 \end{aligned}$$

3.
$$\begin{aligned} x + y - z &= -9 \\ 2x - 3y + 2z &= 13 \\ 3x - 5y - 6z &= -15 \end{aligned}$$

In Exercises 4–6, solve the system using the substitution method.

4.
$$\begin{aligned} x + y + z &= 4 \\ x + y - z &= 4 \\ 3x + 3y + z &= 12 \end{aligned}$$

5.
$$\begin{aligned} 2x + 3y - z &= 9 \\ x - 3y + z &= -6 \\ 3x + y - 4z &= 31 \end{aligned}$$

6.
$$\begin{aligned} x + 2y - 5z &= -12 \\ 2x + 2y - 3z &= -2 \\ 3x - 4y - z &= 11 \end{aligned}$$

7. You found \$6.60 on the ground at school, all in nickels, dimes, and quarters. You have twice as many quarters as dimes and 42 coins in all. How many of each type of coin do you have?

Name ____________________ Date __________

1.4 Notetaking with Vocabulary (continued)

8. Find the values of a, b, and c so that the linear system below has $(3, -2, 1)$ as its only solution. Explain your reasoning.

$$\begin{aligned} 3x + 2y - 7z &= a \\ x + 3y + z &= b \\ 4x - 2y - z &= c \end{aligned}$$

9. Does the system of linear equations have more than one solution? Justify your answer.

$$\begin{aligned} \tfrac{1}{2}x - \tfrac{3}{8}y + \tfrac{1}{8}z &= -\tfrac{5}{4} \\ \tfrac{1}{2}x + \tfrac{1}{4}y + \tfrac{3}{4}z &= 0 \\ -x + 2y - 5z &= 17 \end{aligned}$$

10. If $\angle A$ is three times as large as $\angle B$, and $\angle B$ is 30° smaller than $\angle C$, what are the measures of angles A, B, and C?

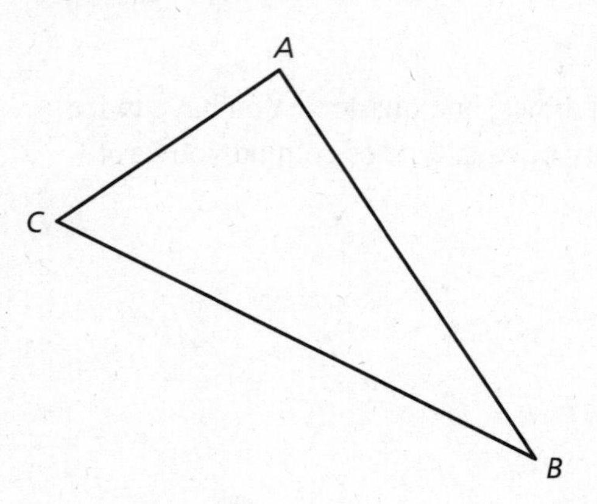

Name______________________________ Date__________

Chapter 2 Maintaining Mathematical Proficiency

Find the *x*-intercept of the graph of the linear equation.

1. $y = 4x + 36$

2. $y = -3x + 5$

3. $y = -10x - 75$

4. $y = 2(x - 9)$

5. $y = -7(x + 10)$

6. $5x + 15y = 60$

Find the distance between the two points.

7. $(1, 3), (-2, 8)$

8. $(-5, 0), (-9, 2)$

9. $(3, 7), (10, 4)$

10. $(6, -2), (-3, 0)$

11. $(9, -1), (9, 8)$

12. $(0, 5), (-4, -6)$

13. A student uses the Distance Formula to find the distance between two points (a, b) and (c, d). What does the step $\sqrt{(c - a)^2 + (0)^2}$ tell the student about the relationship between the two points? How could the student have found the distance between the two points using another method?

Name ______________________________ Date __________

2.1 Transformations of Quadratic Functions
For use with Exploration 2.1

Essential Question How do the constants a, h, and k affect the graph of the quadratic function $g(x) = a(x - h)^2 + k$?

1 EXPLORATION: Identifying Graphs of Quadratic Functions

Work with a partner. Match each quadratic function with its graph. Explain your reasoning. Then use a graphing calculator to verify that your answer is correct.

a. $g(x) = -(x - 2)^2$ **b.** $g(x) = (x - 2)^2 + 2$ **c.** $g(x) = -(x + 2)^2 - 2$

d. $g(x) = 0.5(x - 2)^2 - 2$ **e.** $g(x) = 2(x - 2)^2$ **f.** $g(x) = -(x + 2)^2 + 2$

A.
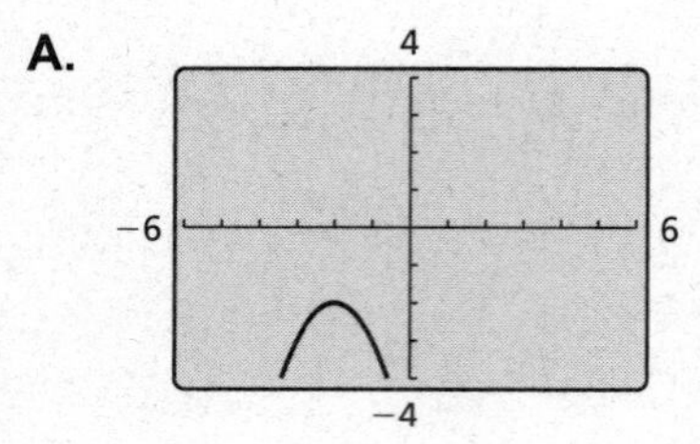

B.
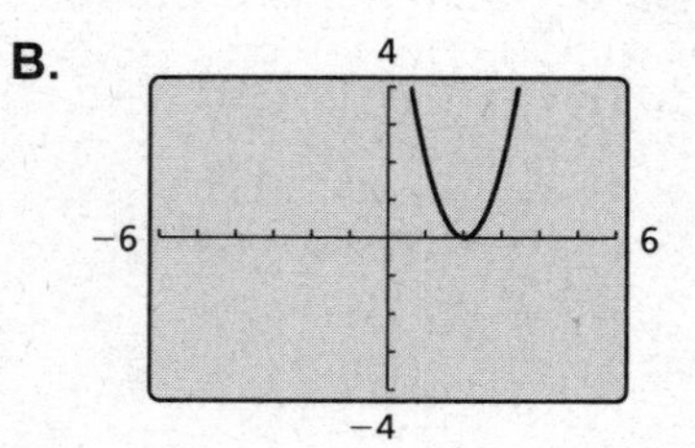

C.
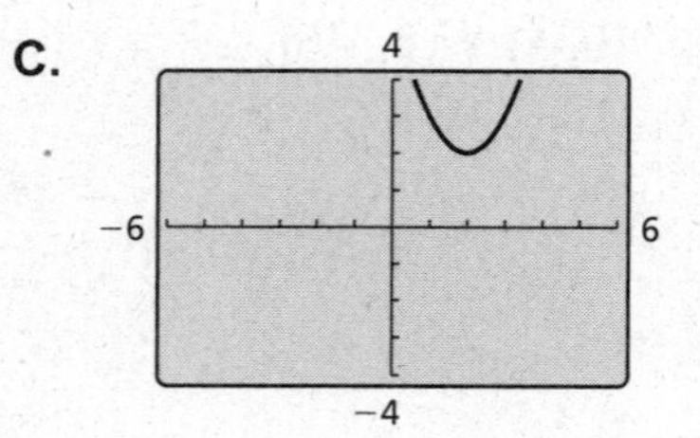

D.
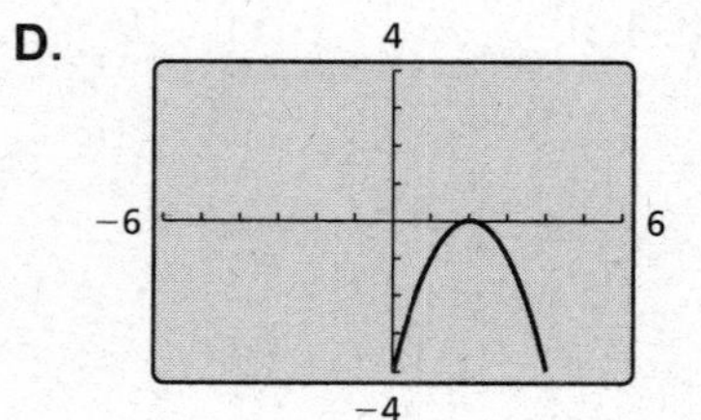

E.
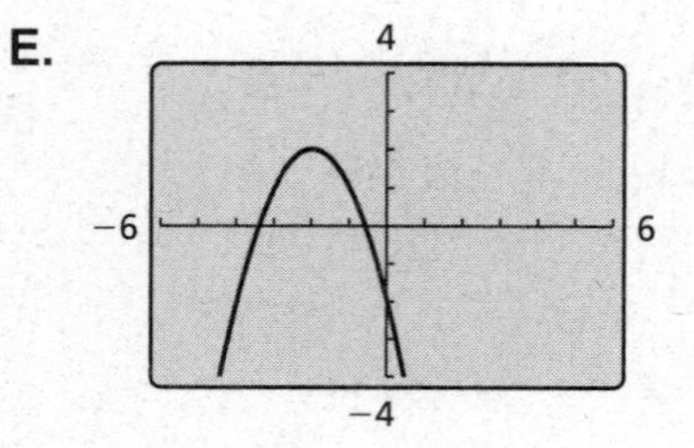

F.
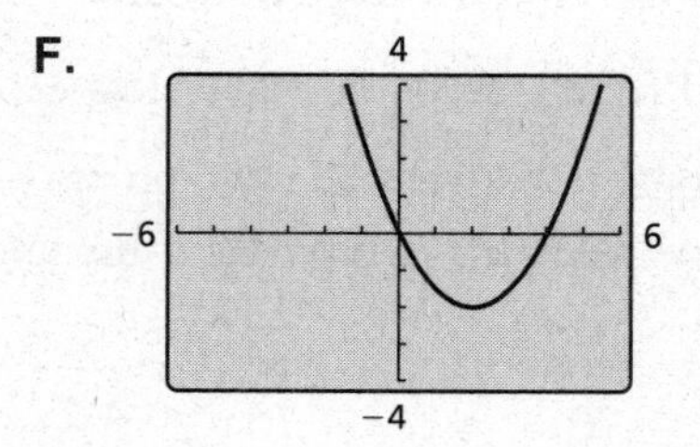

Communicate Your Answer

2. How do the constants a, h, and k affect the graph of the quadratic function $g(x) = a(x - h)^2 + k$?

3. Write the equation of the quadratic function whose graph is shown. Explain your reasoning. Then use a graphing calculator to verify that your equation is correct.

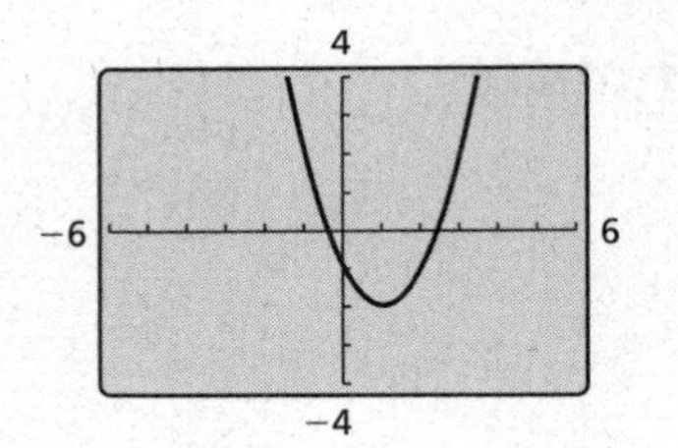

Name ______________________________ Date __________

Notetaking with Vocabulary

For use after Lesson 2.1

In your own words, write the meaning of each vocabulary term.

quadratic function

parabola

vertex of a parabola

vertex form

Core Concepts

Horizontal Translations

$$f(x) = x^2$$

$$f(x - h) = (x - h)^2$$

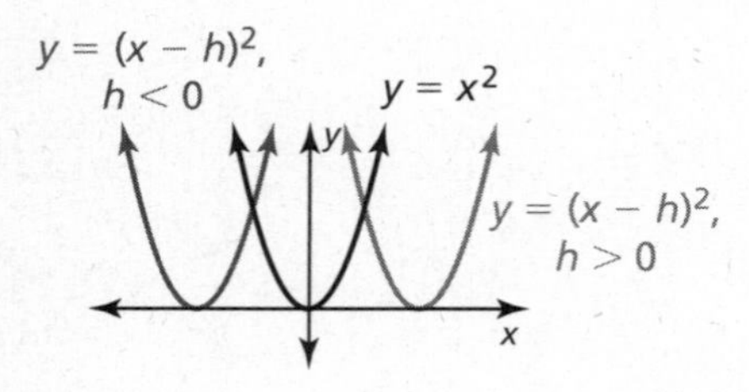

- shifts left when $h < 0$
- shifts right when $h > 0$

Vertical Translations

$$f(x) = x^2$$

$$f(x) + k = x^2 + k$$

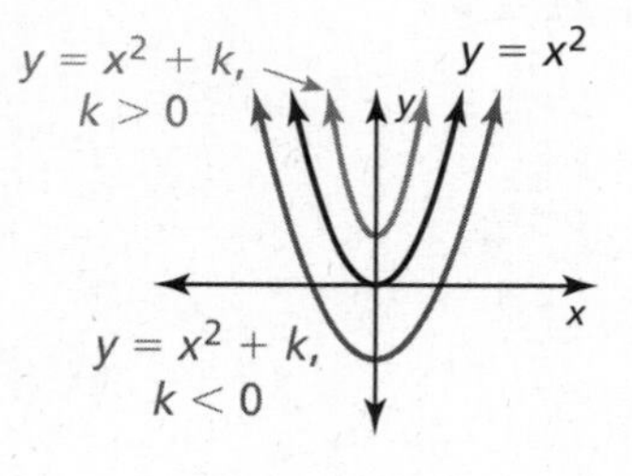

- shifts down when $k < 0$
- shifts up when $k > 0$

Notes:

Name______________________________ Date__________

2.1 Notetaking with Vocabulary (continued)

Reflections in the *x*-Axis

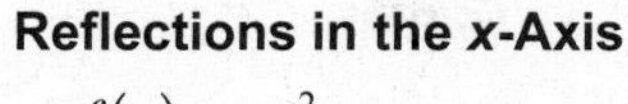

$$f(x) = x^2$$

$$-f(x) = -(x^2) = -x^2$$

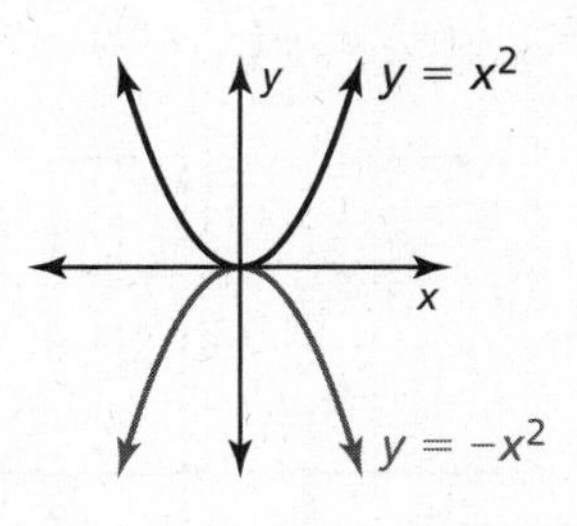

flips over the *x*-axis

Reflections in the *y*-Axis

$$f(x) = x^2$$

$$f(-x) = (-x)^2 = x^2$$

$y = x^2$ is its own reflection in the *y*-axis.

Horizontal Stretches and Shrinks

$$f(x) = x^2$$

$$f(ax) = (ax)^2$$

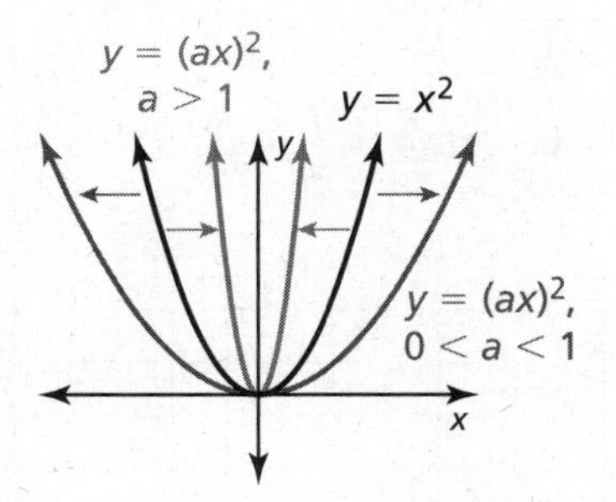

- horizontal stretch (away from *y*-axis) when $0 < a < 1$
- horizontal shrink (toward *y*-axis) when $a > 1$

Vertical Stretches and Shrinks

$$f(x) = x^2$$

$$a \bullet f(x) = ax^2$$

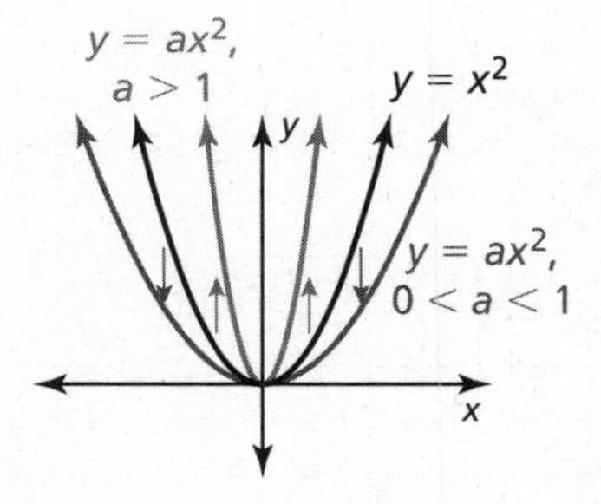

- vertical stretch (away from *x*-axis) when $a > 1$
- vertical shrink (toward *x*-axis) when $0 < a < 1$

Notes:

Extra Practice

In Exercises 1–6, describe the transformation of $f(x) = x^2$ represented by *g*. Then graph the function.

1. $g(x) = x^2 + 4$

2. $g(x) = (x - 1)^2 - 3$

3. $g(x) = -(x + 9)^2$

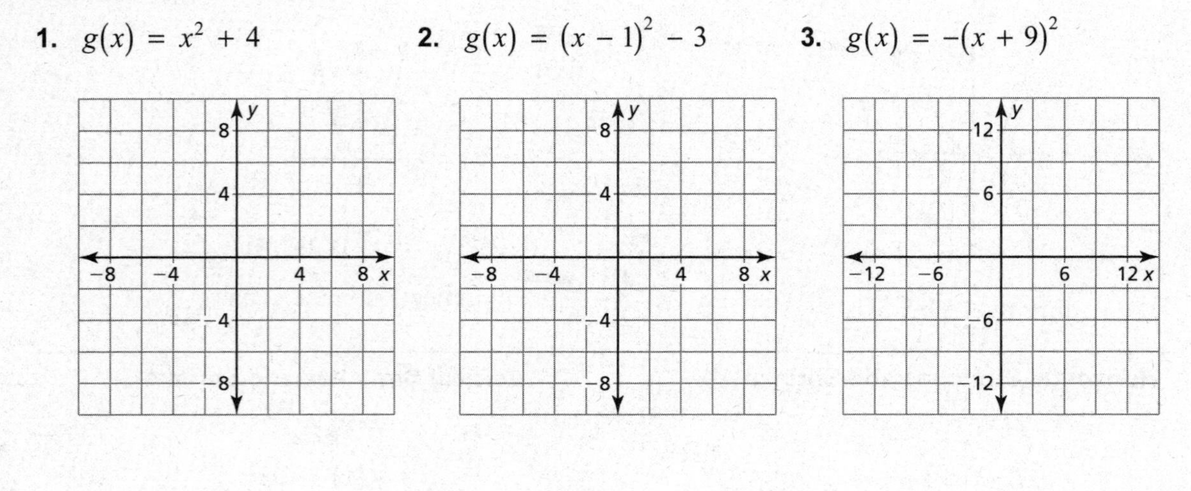

4. $g(x) = x^2 - 7$

5. $g(x) = \frac{1}{3}x^2 - 6$

6. $g(x) = (-4x)^2$

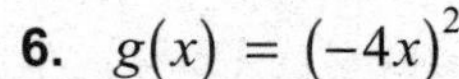

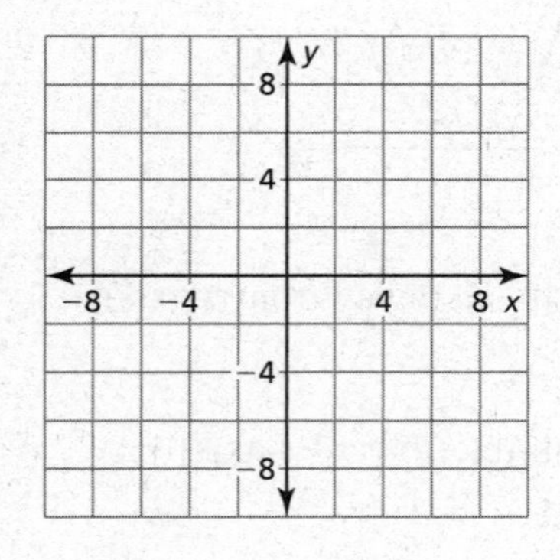

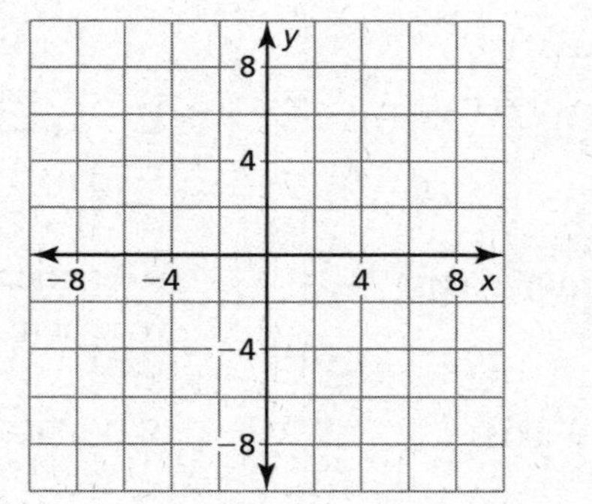

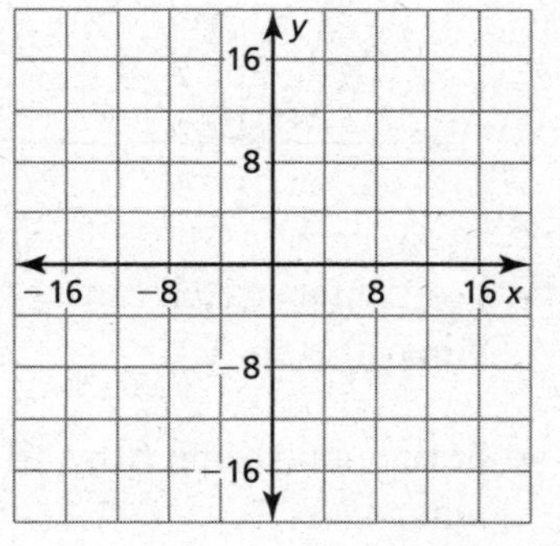

7. Consider the function $f(x) = -10(x - 5)^2 + 7$. Describe the transformation of the graph of the parent quadratic function. Then identify the vertex.

Name____ Date__________

2.2 Characteristics of Quadratic Functions

For use with Exploration 2.2

Essential Question What type of symmetry does the graph of $f(x) = a(x - h)^2 + k$ have and how can you describe this symmetry?

1 EXPLORATION: Parabolas and Symmetry

Work with a partner.

a. Complete the table. Then use the values in the table to sketch the graph of the function $f(x) = \frac{1}{2}x^2 - 2x - 2$ on graph paper.

x	−2	−1	0	1	2
f(x)					

x	3	4	5	6
f(x)				

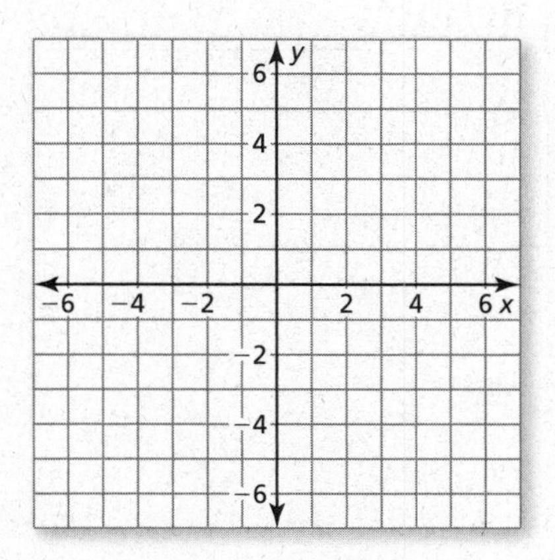

b. Use the results in part (a) to identify the vertex of the parabola.

c. Find a vertical line on your graph paper so that when you fold the paper, the left portion of the graph coincides with the right portion of the graph. What is the equation of this line? How does it relate to the vertex?

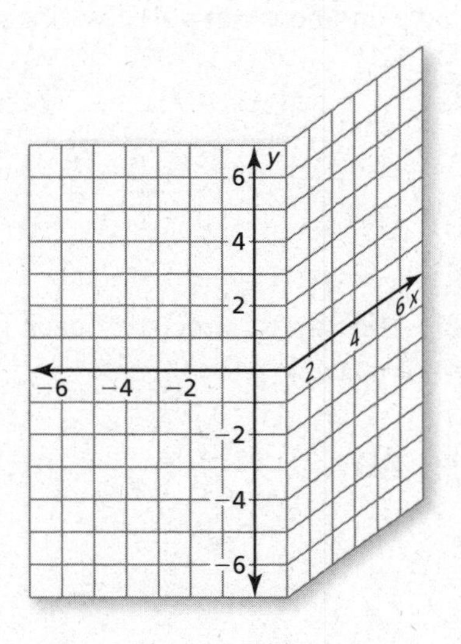

d. Show that the vertex form $f(x) = \frac{1}{2}(x - 2)^2 - 4$ is equivalent to the function given in part (a).

2 EXPLORATION: Parabolas and Symmetry

Work with a partner. Repeat Exploration 1 for the function given by $f(x) = -\frac{1}{3}x^2 + 2x + 3 = -\frac{1}{3}(x - 3)^2 + 6$.

x	−2	−1	0	1	2
f(x)					

x	3	4	5	6
f(x)				

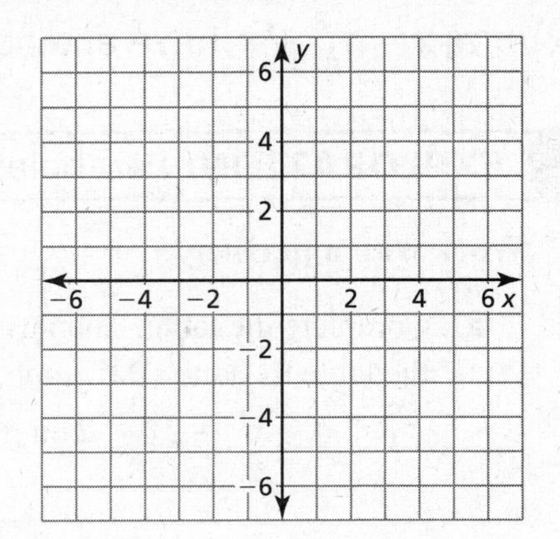

Communicate Your Answer

3. What type of symmetry does the graph of the parabola $f(x) = a(x - h)^2 + k$ have and how can you describe this symmetry?

4. Describe the symmetry of each graph. Then use a graphing calculator to verify your answer.

a. $f(x) = -(x - 1)^2 + 4$ **b.** $f(x) = (x + 1)^2 - 2$ **c.** $f(x) = 2(x - 3)^2 + 1$

d. $f(x) = \frac{1}{2}(x + 2)^2$ **e.** $f(x) = -2x^2 + 3$ **f.** $f(x) = 3(x - 5)^2 + 2$

Name______________________________ Date__________

2.2 Notetaking with Vocabulary
For use after Lesson 2.2

In your own words, write the meaning of each vocabulary term.

axis of symmetry

standard form

minimum value

maximum value

intercept form

Core Concepts

Properties of the graph of $f(x) = ax^2 + bx + c$

$y = ax^2 + bx + c, a > 0$

$y = ax^2 + bx + c, a < 0$

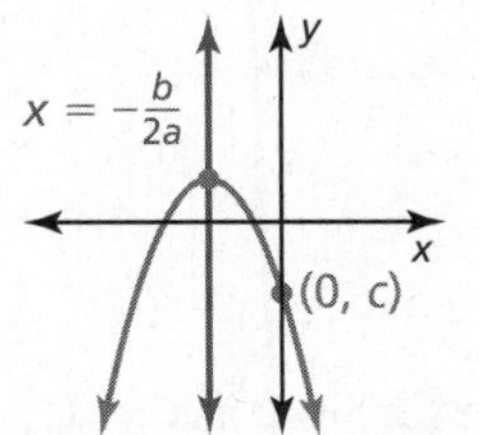

- The parabola opens up when $a > 0$ and open down when $a < 0$.
- The graph is narrower than the graph of $f(x) = x^2$ when $|a| > 1$ and wider when $|a| < 1$.
- The axis of symmetry is $x = -\dfrac{b}{2a}$ and the vertex is $\left(-\dfrac{b}{2a}, f\left(-\dfrac{b}{2a}\right)\right)$.
- The y-intercept is c. So, the point $(0, c)$ is on the parabola.

Notes:

Minimum and Maximum Values

For the quadratic function $f(x) = ax^2 + bx + c$, the y-coordinate of the vertex is the **minimum value** of the function when $a > 0$ and the **maximum value** when $a < 0$.

$a > 0$

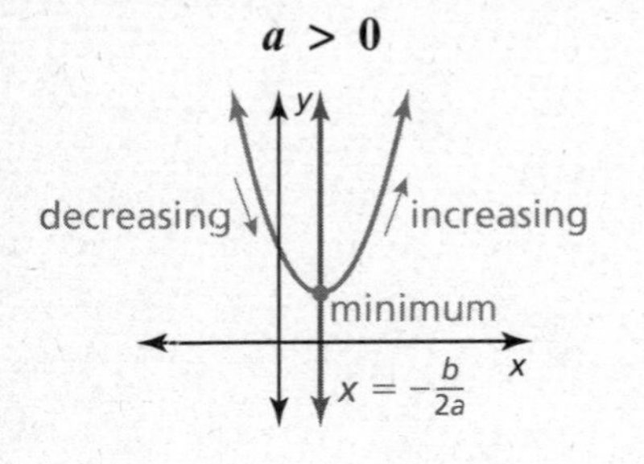

$a < 0$

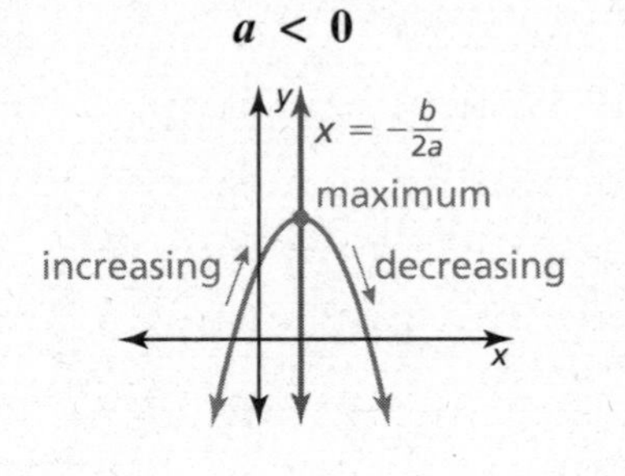

- Minimum value: $f\left(-\dfrac{b}{2a}\right)$
- Domain: All real numbers
- Range: $y \geq f\left(-\dfrac{b}{2a}\right)$
- Decreasing to the left of $x = -\dfrac{b}{2a}$
- Increasing to the right of $x = -\dfrac{b}{2a}$

- Maximum value: $f\left(-\dfrac{b}{2a}\right)$
- Domain: All real numbers
- Range: $y \leq f\left(-\dfrac{b}{2a}\right)$
- Increasing to the left of $x = -\dfrac{b}{2a}$
- Decreasing to the right of $x = -\dfrac{b}{2a}$

Notes:

Properties of the graph of $f(x) = a(x - p)(x - q)$

- Because $f(p) = 0$ and $f(q) = 0$, p and q are the x-intercepts of the graph of the function.
- The axis of symmetry is halfway between $(p, 0)$ and $(q, 0)$.

  So, the axis of symmetry is $x = \dfrac{p + q}{2}$.
- The parabola opens up when $a > 0$ and opens down when $a < 0$.

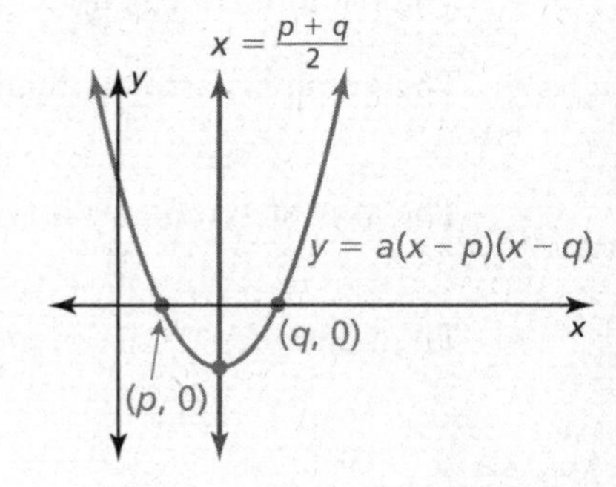

Notes:

Name______________________________ Date__________

2.2 Notetaking with Vocabulary (continued)

Extra Practice

In Exercises 1–3, graph the function. Label the vertex and axis of symmetry. Find the minimum or maximum value of the function. Describe the domain and range of the function, and where the function is increasing and decreasing.

1. $f(x) = (x + 1)^2$

2. $y = -2(x - 4)^2 - 5$

3. $t(x) = \frac{3}{2}x^2 - 3x - 1$

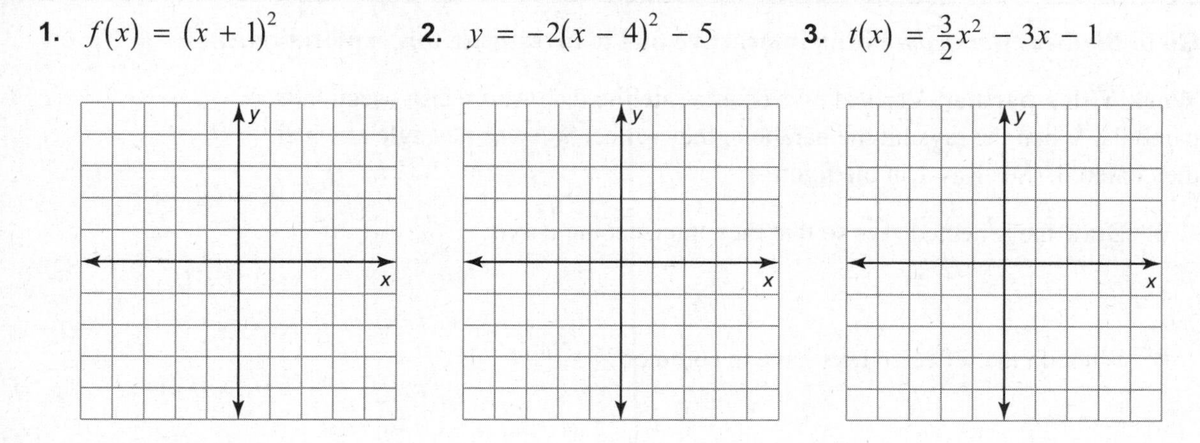

In Exercises 4 and 5, graph the function. Label the *x*-intercept(s), vertex, and axis of symmetry.

4. $f(x) = 4(x + 4)(x - 3)$

5. $f(x) = -7x(x - 6)$

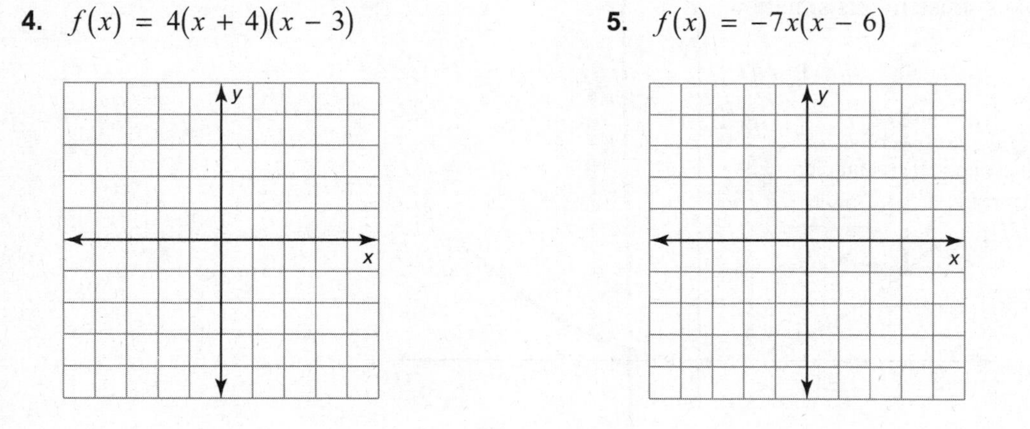

6. A softball player hits a ball whose path is modeled by $f(x) = -0.0005x^2 + 0.2127x + 3$, where x is the distance from home plate (in feet) and y is the height of the ball above the ground (in feet). What is the highest point this ball will reach? If the ball was hit to center field which has an 8 foot fence located 410 feet from home plate, was this hit a home run? Explain.

Name ______________________________ Date ________

2.3 Focus of a Parabola

For use with Exploration 2.3

Essential Question What is the focus of a parabola?

1 EXPLORATION: Analyzing Satellite Dishes

Go to *BigIdeasMath.com* for an interactive tool to investigate this exploration.

Work with a partner. Vertical rays enter a satellite dish whose cross section is a parabola. When the rays hit the parabola, they reflect at the same angle at which they entered. (See Ray 1 in the figure.)

a. Draw the reflected rays so that they intersect the *y*-axis.

b. What do the reflected rays have in common?

c. The optimal location for the receiver of the satellite dish is at a point called the *focus* of the parabola. Determine the location of the focus. Explain why this makes sense in this situation.

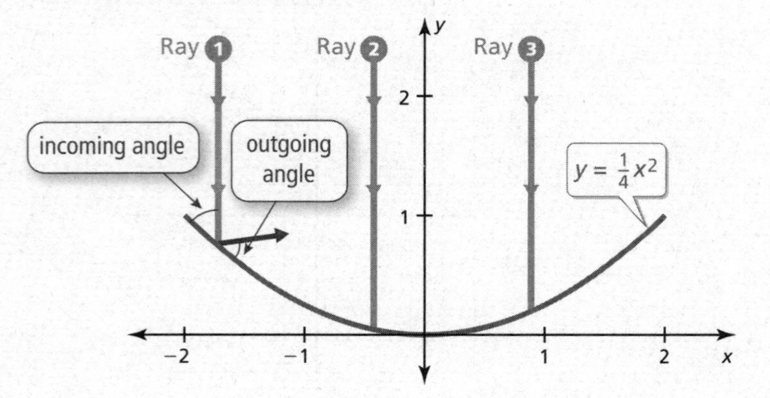

Name______________________________ Date__________

2 EXPLORATION: Analyzing Spotlights

Go to *BigIdeasMath.com* for an interactive tool to investigate this exploration.

Work with a partner. Beams of light are coming from the bulb in a spotlight, located at the focus of the parabola. When the beams hit the parabola, they reflect at the same angle at which they hit. (See Beam 1 in the figure.) Draw the reflected beams. What do they have in common? Would you consider this to be the optimal result? Explain.

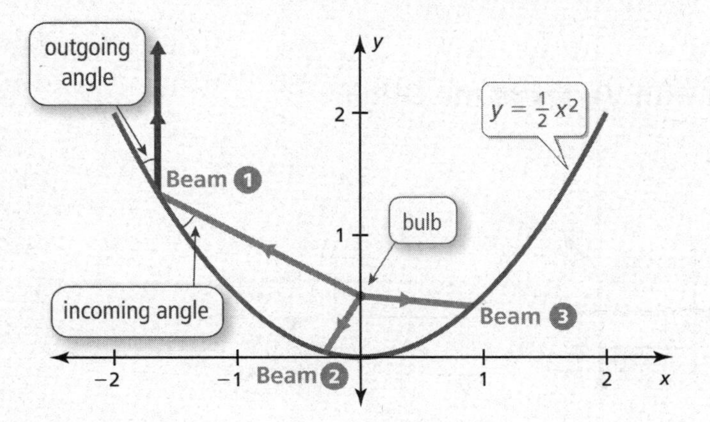

Communicate Your Answer

3. What is the focus of a parabola?

4. Describe some of the properties of the focus of a parabola.

Name ______________________________ Date __________

2.3 Notetaking with Vocabulary

For use after Lesson 2.3

In your own words, write the meaning of each vocabulary term.

focus

directrix

Core Concepts

Standard Equations of a Parabola with Vertex at the Origin

Vertical axis of symmetry $(x = 0)$

Equation: $y = \frac{1}{4p}x^2$

Focus: $(0, p)$

Directrix: $y = -p$

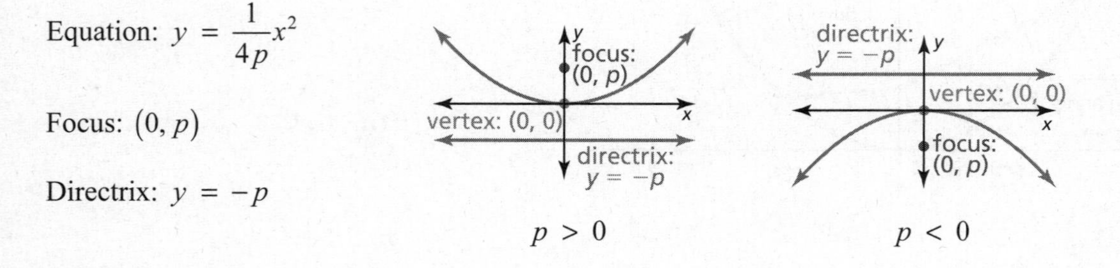

$p > 0$ $p < 0$

Horizontal axis of symmetry $(y = 0)$

Equation: $x = \frac{1}{4p}y^2$

Focus: $(p, 0)$

Directrix: $x = -p$

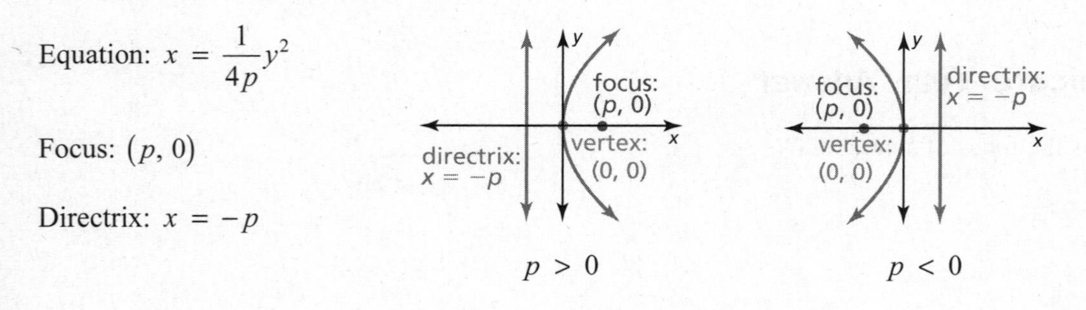

$p > 0$ $p < 0$

Notes:

Standard Equations of a Parabola with Vertex at (h, k)

Vertical axis of symmetry $(x = h)$

Equation: $y = \frac{1}{4p}(x - h)^2 + k$

Focus: $(h, k + p)$

Directrix: $y = k - p$

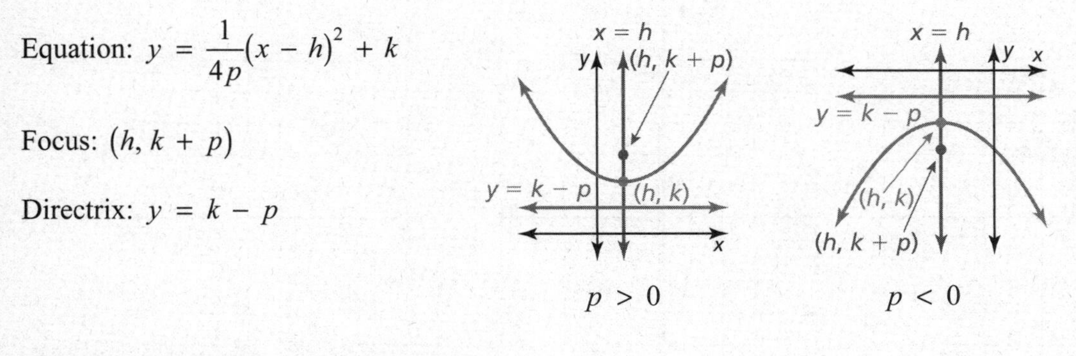

Horizontal axis of symmetry $(y = k)$

Equation: $x = \frac{1}{4p}(y - k)^2 + h$

Focus: $(h + p, k)$

Directrix: $x = h - p$

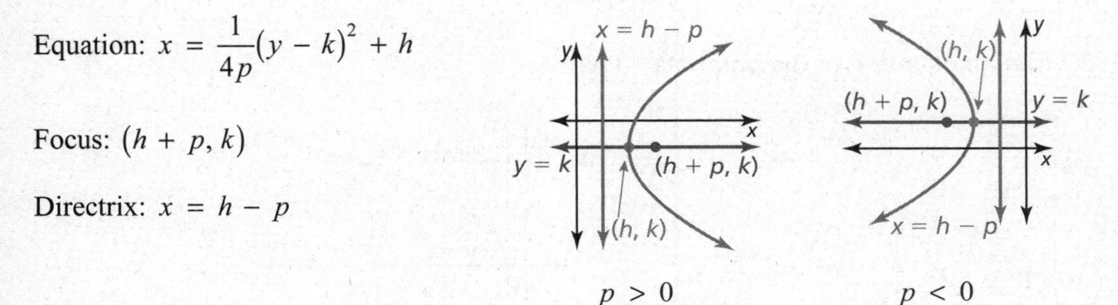

Notes:

Extra Practice

In Exercises 1 and 2, use the Distance Formula to write an equation of the parabola.

1. focus: $(0, -8)$ directrix: $y = 8$

2. vertex: $(0, 0)$ focus: $(0, 1)$

2.3 Notetaking with Vocabulary (continued)

In Exercises 3–5, identify the focus, directrix, and axis of symmetry of the parabola. Graph the equation.

3. $x^2 = -2y$ **4.** $-5x + \frac{1}{3}y^2 = 0$ **5.** $y = -2(x + 1)^2 - 3$

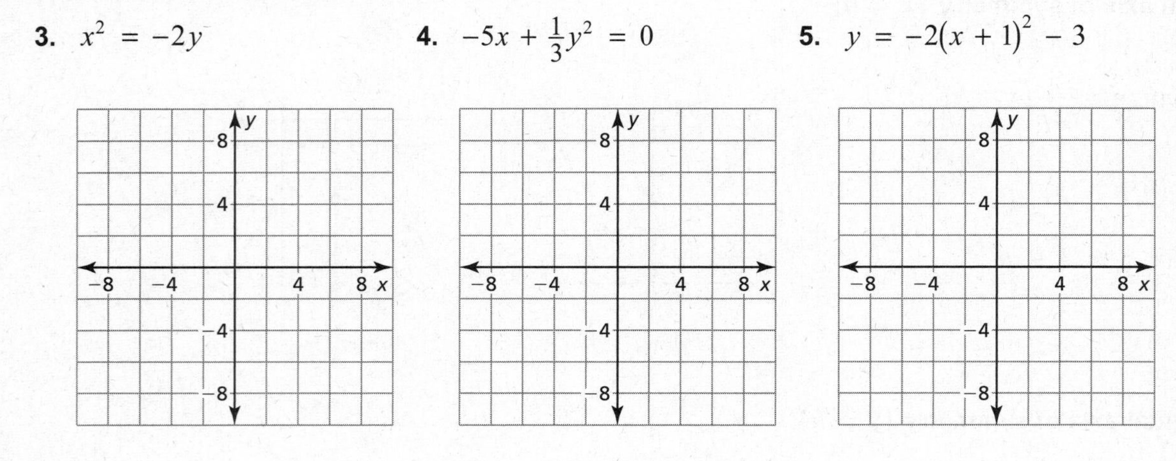

In Exercises 6–8, write an equation of the parabola shown.

6.

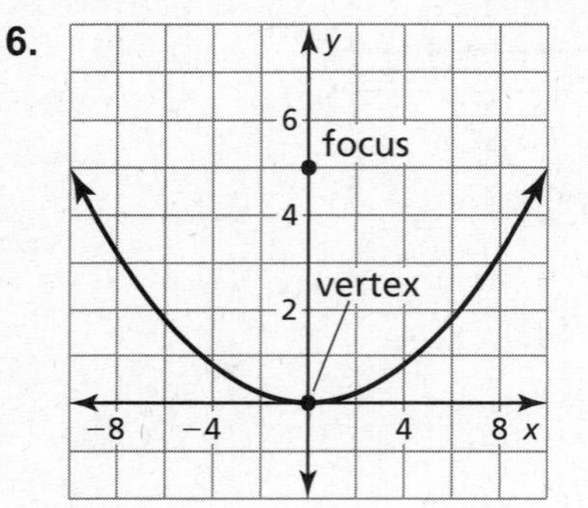

7.

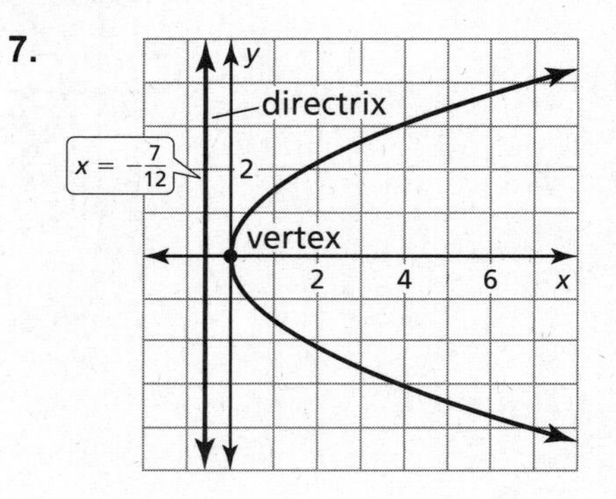

8.

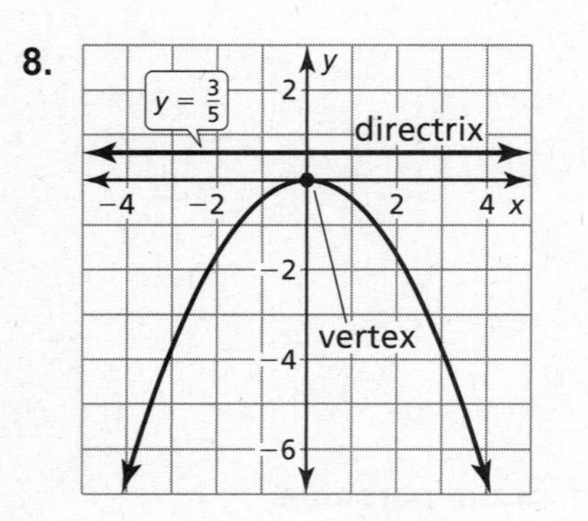

9. The cross section of a parabolic sound reflector at the Olympics has a diameter of 20 inches and is 25 inches deep. Write an equation that represents the cross section of the reflector with its vertex at $(0, 0)$ and its focus to the left of the vertex.

Name____ Date__________

2.4 Modeling with Quadratic Functions

For use with Exploration 2.4

Essential Question How can you use a quadratic function to model a real-life situation?

1 EXPLORATION: Modeling with a Quadratic Function

Work with a partner. The graph shows a quadratic function of the form $P(t) = at^2 + bt + c$ which approximates the yearly profits for a company, where $P(t)$ is the profit in year t.

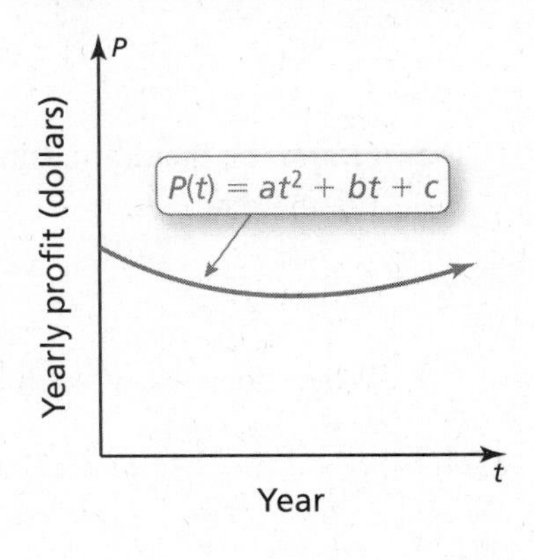

a. Is the value of a positive, negative, or zero? Explain.

b. Write an expression in terms of a and b that represents the year t when the company made the least profit.

c. The company made the same yearly profits in 2004 and 2012. Estimate the year in which the company made the least profit.

d. Assume that the model is still valid today. Are the yearly profits currently increasing, decreasing, or constant? Explain.

2 EXPLORATION: Modeling with a Graphing Calculator

Go to *BigIdeasMath.com* for an interactive tool to investigate this exploration.

Work with a partner. The table shows the heights h (in feet) of a wrench t seconds after it has been dropped from a building under construction.

Time, t	0	1	2	3	4
Height, h	400	384	336	256	144

a. Use a graphing calculator to create a scatter plot of the data, as shown at the right. Explain why the data appear to fit a quadratic model.

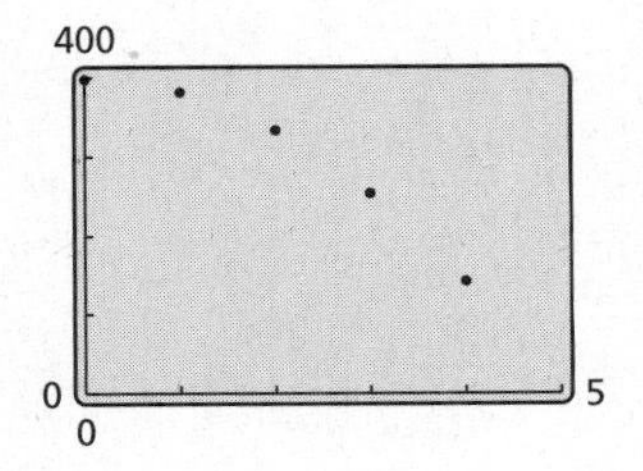

Name ____ Date __________

2.4 Modeling with Quadratic Functions (continued)

2 EXPLORATION: Modeling with a Graphing Calculator (continued)

b. Use the *quadratic regression* feature to find a quadratic model for the data.

c. Graph the quadratic function on the same screen as the scatter plot to verify that it fits the data.

d. When does the wrench hit the ground? Explain.

Communicate Your Answer

3. How can you use a quadratic function to model a real-life situation?

4. Use the Internet or some other reference to find examples of real-life situations that can be modeled by quadratic functions.

Name____ Date__________

2.4 Notetaking with Vocabulary

For use after Lesson 2.4

In your own words, write the meaning of each vocabulary term.

average rate of change

system of three linear equations

Core Concepts

Writing Quadratic Equations

Given a point and the vertex (h, k)	Use vertex form: $y = a(x - h)^2 + k$
Given a point and x-intercepts p and q	Use intercept form: $y = a(x - p)(x - q)$
Given three points	Write and solve a system of three equations in three variables.

Notes:

Name ____ Date __________

Extra Practice

In Exercises 1–4, write an equation of the parabola in vertex form.

1. 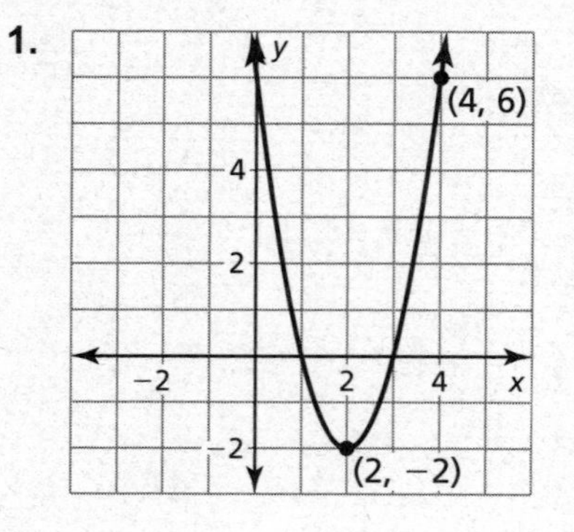

2. 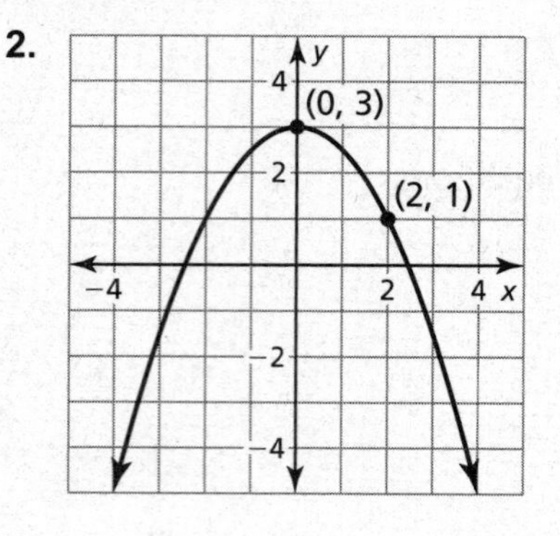

3. passes through $(-3, 0)$ and has vertex $(-1, -8)$

4. passes through $(-4, 7)$ and has vertex $(-2, 5)$

In Exercises 5–8, write an equation of the parabola in intercept form.

5. 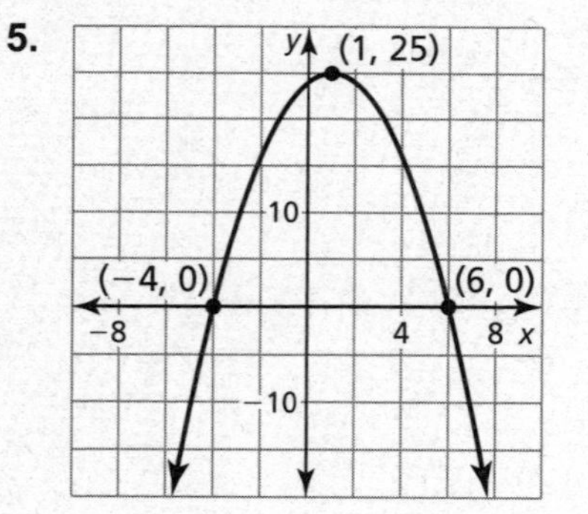

6. 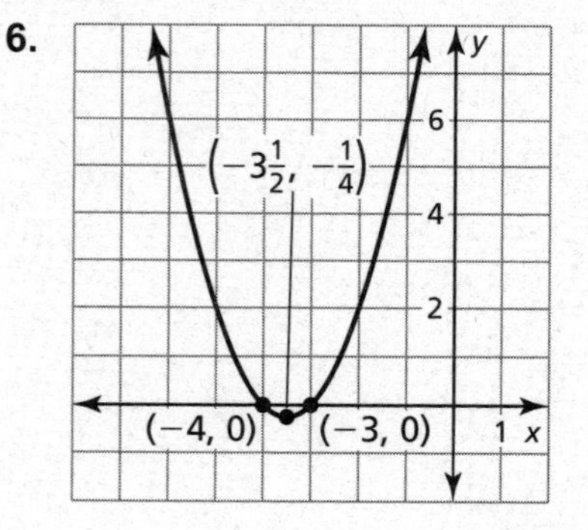

7. x-intercepts of -5 and 8; passes through $(1, 84)$

8. x-intercepts of 7 and 10; passes through $(-2, 27)$

In Exercises 9–11, analyze the differences in the outputs to determine whether the data are *linear*, *quadratic* or *neither*. If linear or quadratic, write an equation that fits the data.

9.

Time (seconds), *x*	1	2	3	4	5	6
Distance (feet), *y*	424	416	376	304	200	64

10.

Time (days), *x*	0	3	6	9	12	15
Height (inches), *y*	36	30	24	18	12	6

11.

Time (years), *x*	1	2	3	4	5	6
Profit (dollars), *y*	5	15	45	135	405	1215

12. The table shows a university's budget (in millions of dollars) over a 10-year period, where $x = 0$ represents the first year in the 10-year period.

Years, *x*	0	1	2	3	4	5	6	7	8	9
Budget, *y*	65	32	22	40	65	92	114	128	140	150

a. Use a graphing calculator to create a scatter plot. Which better represents the data, a line or a parabola? Explain.

b. Use the *regression* feature of your calculator to find the model that best fits the data.

c. Use the model in part (b) to predict when the budget of the university is $500,000,000.00.

Name ______________________ Date __________

Chapter 3 Maintaining Mathematical Proficiency

Simplify the expression.

1. $\sqrt{50}$

2. $-\sqrt{96}$

3. $\sqrt{\frac{3}{121}}$

4. $\sqrt{200}$

5. $\sqrt{\frac{75}{81}}$

6. $-\sqrt{\frac{14}{144}}$

7. $-\sqrt{54}$

8. $\sqrt{250}$

Factor the polynomial.

9. $x^2 - 100$

10. $4x^2 - 49$

11. $16x^2 - 9$

12. $x^2 - 30x + 225$

13. $x^2 + 16x + 64$

14. $25x^2 + 10x + 1$

15. Explain why the expression $81 - x^4$ *cannot* be factored into $(3 + x)^2(3 - x)^2$.

Name____ Date__________

3.1 Solving Quadratic Equations

For use with Exploration 3.1

Essential Question How can you use the graph of a quadratic equation to determine the number of real solutions of the equation?

1 EXPLORATION: Matching a Quadratic Function with Its Graph

Work with a partner. Match each quadratic function with its graph. Explain your reasoning. Determine the number of x-intercepts of the graph.

a. $f(x) = x^2 - 2x$ **b.** $f(x) = x^2 - 2x + 1$ **c.** $f(x) = x^2 - 2x + 2$

d. $f(x) = -x^2 + 2x$ **e.** $f(x) = -x^2 + 2x - 1$ **f.** $f(x) = -x^2 + 2x - 2$

A.

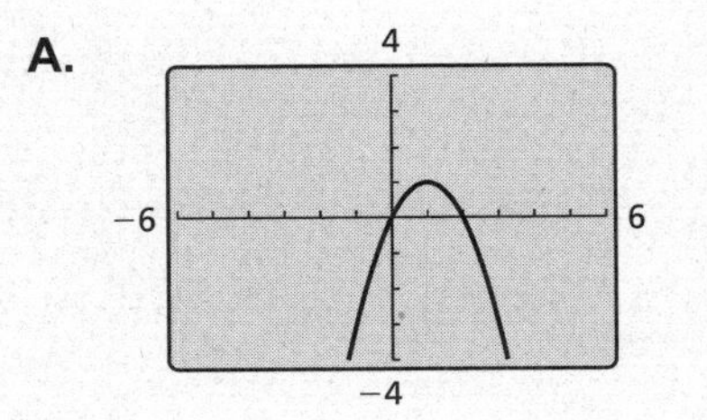

B.

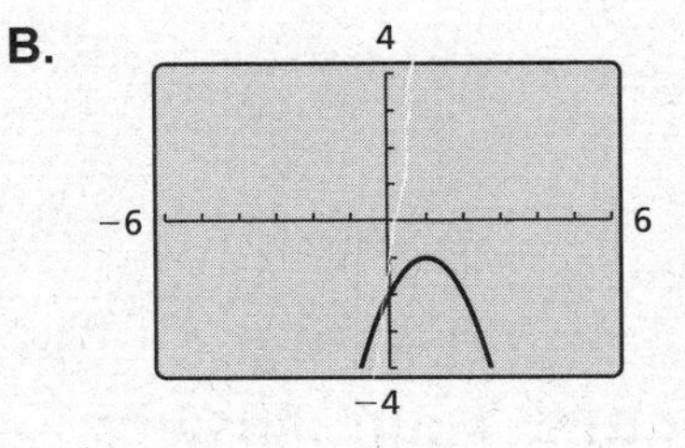

C.

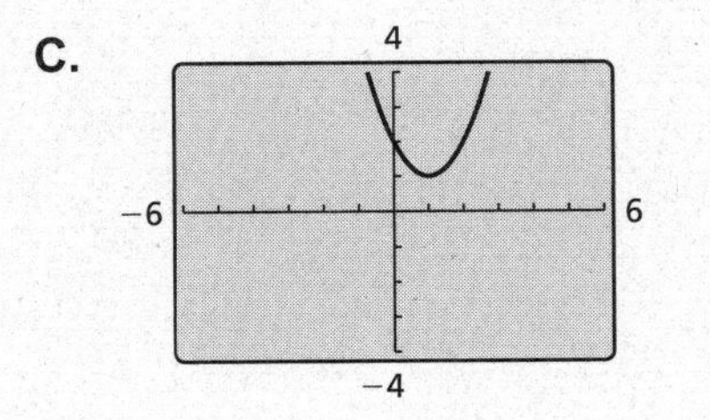

D.

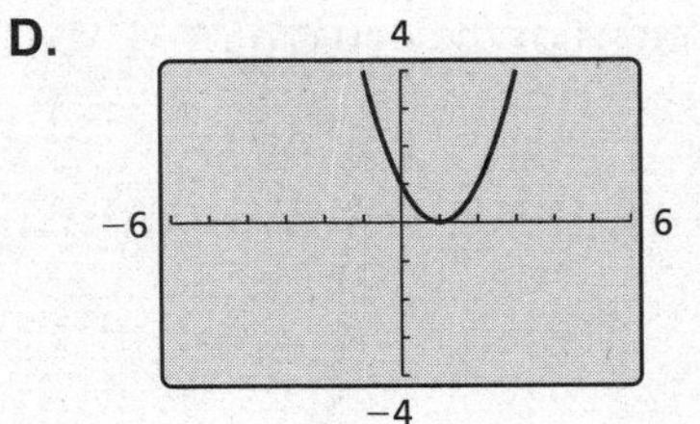

E.

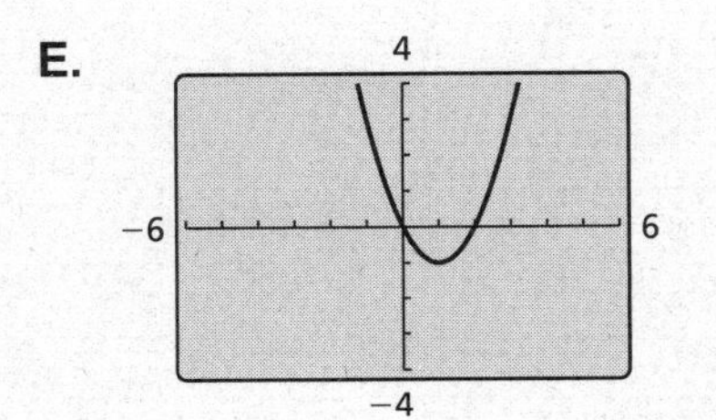

F.

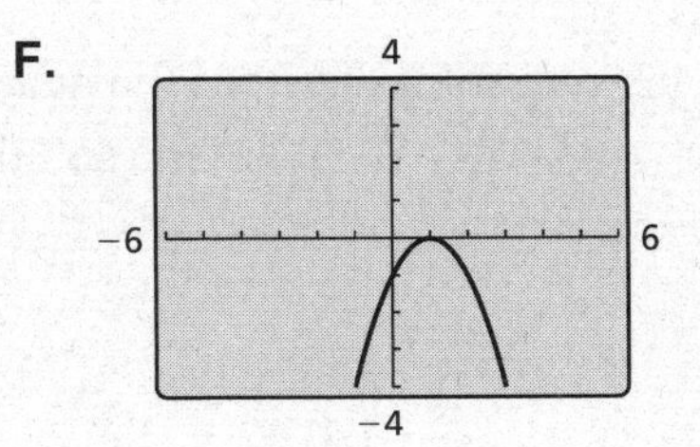

Name ______________________________ Date __________

3.1 Solving Quadratic Equations (continued)

2 EXPLORATION: Solving Quadratic Equations

Work with a partner. Use the results of Exploration 1 to find the real solutions (if any) of each quadratic equation.

a. $x^2 - 2x = 0$

b. $x^2 - 2x + 1 = 0$

c. $x^2 - 2x + 2 = 0$

d. $-x^2 + 2x = 0$

e. $-x^2 + 2x - 1 = 0$

f. $-x^2 + 2x - 2 = 0$

Communicate Your Answer

3. How can you use the graph of a quadratic equation to determine the number of real solutions of the equation?

4. How many real solutions does the quadratic equation $x^2 + 3x + 2 = 0$ have? How do you know? What are the solutions?

Name____ Date__________

3.1 Notetaking with Vocabulary
For use after Lesson 3.1

In your own words, write the meaning of each vocabulary term.

quadratic equation in one variable

root of an equation

zero of a function

Core Concepts

Solving Quadratic Equations

By graphing Find the x-intercepts of the related function $y = ax^2 + bx + c$.

Using square roots Write the equation in the form $u^2 = d$, where u is an algebraic expression, and solve by taking the square root of each side.

By factoring Write the polynomial equation $ax^2 + bx + c = 0$ in factored form and solve using the Zero-Product Property.

Notes:

Name ______________________________ Date ________

3.1 Notetaking with Vocabulary (continued)

Zero-Product Property

Words If the product of two expressions is zero, then one or both of the expressions equal zero.

Algebra If A and B are expressions and $AB = 0$, then $A = 0$ or $B = 0$.

Notes:

Extra Practice

In Exercises 1–3, solve the equation by graphing.

1. $x^2 - 11x + 24 = 0$

2. $13 = -x^2 - 12$

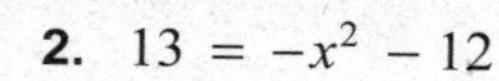

3. $12x^2 = 5x + 2$

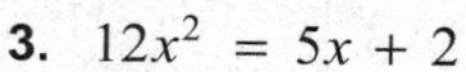

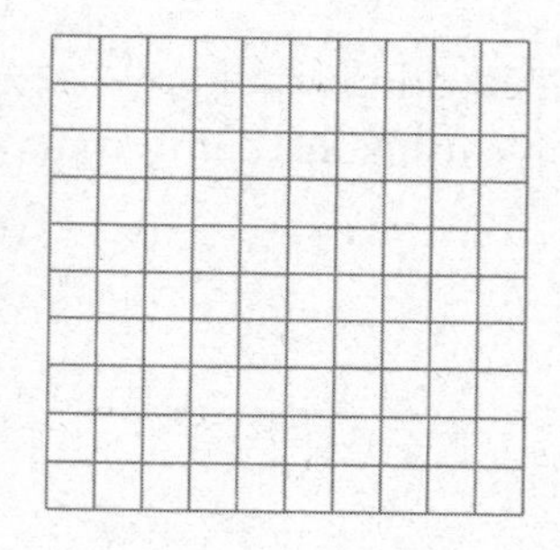

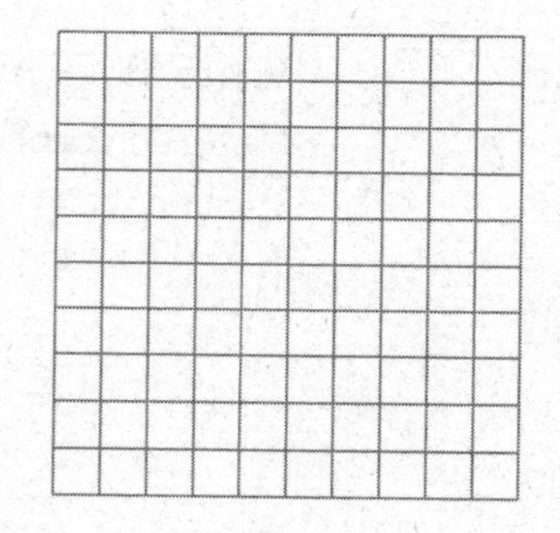

In Exercises 4–6, solve the equation using square roots.

4. $t^2 = 400$

5. $(2k + 3)^2 - 19 = 81$

6. $\frac{1}{7}p^2 = \frac{5}{7}p^2 - 20$

Name______________________________ Date__________

3.1 Notetaking with Vocabulary (continued)

In Exercises 7–9, solve the equation by factoring.

7. $0 = x^2 - 12x + 36$

8. $x^2 = 14x - 40$

9. $5x^2 + 5x - 1 = -x^2 + 4x$

10. Which equations have roots that are equivalent to the x-intercepts of the graph shown?

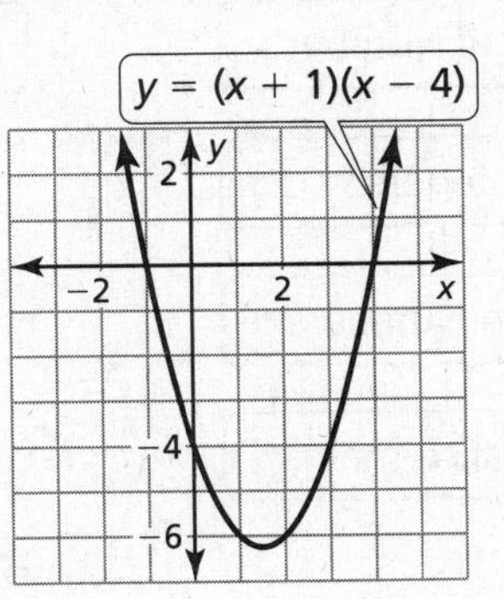

A. $-2x^2 - 10x - 8 = 0$

B. $x^2 - 3x = 4$

C. $(x - 1)(x + 4) = 0$

D. $(x - 1)^2 + 4 = 0$

E. $6x^2 = 18x + 24$

11. A skydiver drops out of an airplane that is flying at an altitude of 4624 feet.

a. Use the formula $h = -16t^2 + h_0$ to write an equation that gives the skydiver's height h (in feet) during free fall t seconds after the skydiver drops out of the airplane.

b. It is possible for the skydiver to wait 18 seconds before pulling the parachute cord? Explain.

Name ______________________________ Date __________

3.2 Complex Numbers
For use with Exploration 3.2

Essential Question What are the subsets of the set of complex numbers?

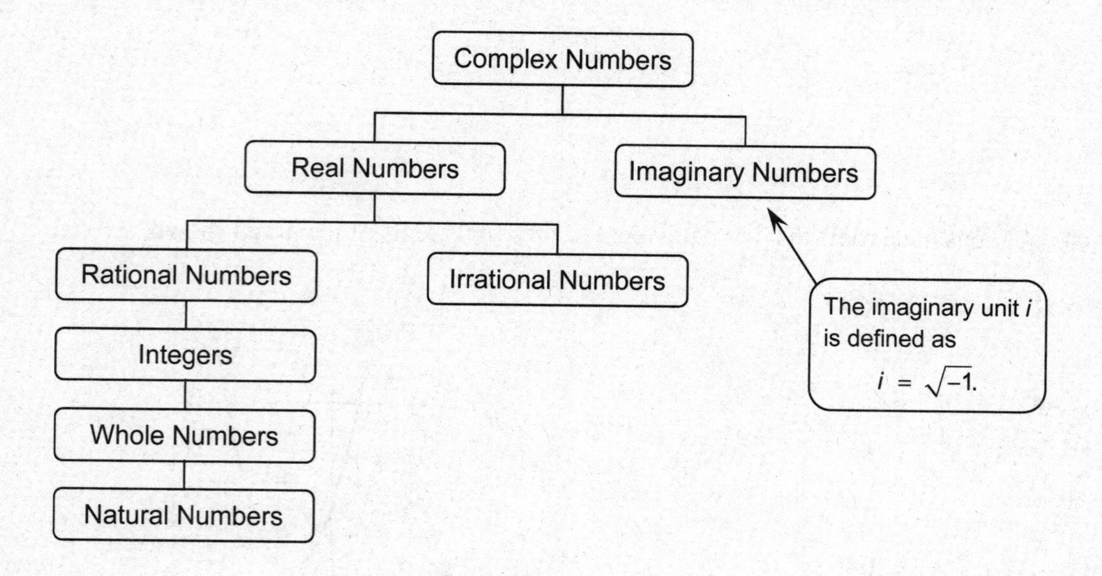

1 EXPLORATION: Classifying Numbers

Work with a partner. Determine which subsets of the set of complex numbers contain each number.

a. $\sqrt{9}$

b. $\sqrt{0}$

c. $-\sqrt{4}$

d. $\sqrt{\frac{4}{9}}$

e. $\sqrt{2}$

f. $\sqrt{-1}$

3.2 Complex Numbers (continued)

2 EXPLORATION: Complex Solutions of Quadratic Equations

Work with a partner. Use the definition of the imaginary unit i to match each quadratic equation with its complex solution. Justify your answers.

a. $x^2 - 4 = 0$

b. $x^2 + 1 = 0$

c. $x^2 - 1 = 0$

d. $x^2 + 4 = 0$

e. $x^2 - 9 = 0$

f. $x^2 + 9 = 0$

A. i

B. $3i$

C. 3

D. $2i$

E. 1

F. 2

Communicate Your Answer

3. What are the subsets of the set of complex numbers? Give an example of a number in each subset.

4. Is it possible for a number to be both whole and natural? natural and rational? rational and irrational? real and imaginary? Explain your reasoning.

Name ______________________ Date __________

3.2 Notetaking with Vocabulary
For use after Lesson 3.2

In your own words, write the meaning of each vocabulary term.

imaginary unit i

complex number

imaginary number

pure imaginary number

Core Concepts

The Square Root of a Negative Number

Property	Example
1. If r is a positive real number, then $\sqrt{-r} = i\sqrt{r}$.	$\sqrt{-3} = i\sqrt{3}$
2. By the first property, it follows that $\left(i\sqrt{r}\right)^2 = -r$.	$\left(i\sqrt{3}\right)^2 = i^2 \bullet 3 = -3$

Notes:

Name________________________________ Date__________

3.2 Notetaking with Vocabulary (continued)

Sums and Differences of Complex Numbers

To add (or subtract) two complex numbers, add (or subtract) their real parts and their imaginary parts separately.

Sum of complex numbers: $(a + bi) + (c + di) = (a + c) + (b + d)i$

Difference of complex numbers: $(a + bi) - (c + di) = (a - c) + (b - d)i$

Notes:

Extra Practice

In Exercises 1–6, find the square root of the number.

1. $\sqrt{-49}$

2. $\sqrt{-4}$

3. $\sqrt{-45}$

4. $-2\sqrt{-100}$

5. $6\sqrt{-121}$

6. $5\sqrt{-75}$

In Exercises 7 and 8, find the values of *x* and *y* that satisfy the equation.

7. $-10x + i = 30 - yi$

8. $44 - \frac{1}{2}yi = -\frac{1}{4}x - 7i$

Name ______________________________ Date __________

3.2 Notetaking with Vocabulary (continued)

In Exercises 9–14, simplify the expression. Then classify the result as a *real number* or *imaginary number*. If the result is an *imaginary number*, specify if it is a *pure imaginary number*.

9. $(-8 + 3i) + (-1 - 2i)$

10. $(36 - 3i) - (12 + 24i)$

11. $(16 + i) + (-16 - 8i)$

12. $(-5 - 5i) - (-6 - 6i)$

13. $(-1 + 9i)(15 - i)$

14. $(13 + i)(13 - i)$

15. Find the impedance of the series circuit.

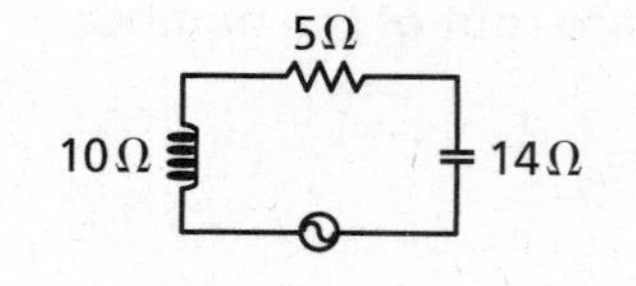

In Exercises 16–18, solve the equation. Check your solution(s).

16. $0 = 5x^2 + 25$

17. $x^2 - 10 = -18$

18. $-\frac{1}{3}x^2 = \frac{1}{5} + \frac{4}{3}x^2$

19. Sketch a graph of a function that has two real zeros at −2 and 2. Then sketch a graph on the same grid of a function that has two imaginary zeros of $-2i$ and $2i$. Explain the difference in the graphs of the two functions.

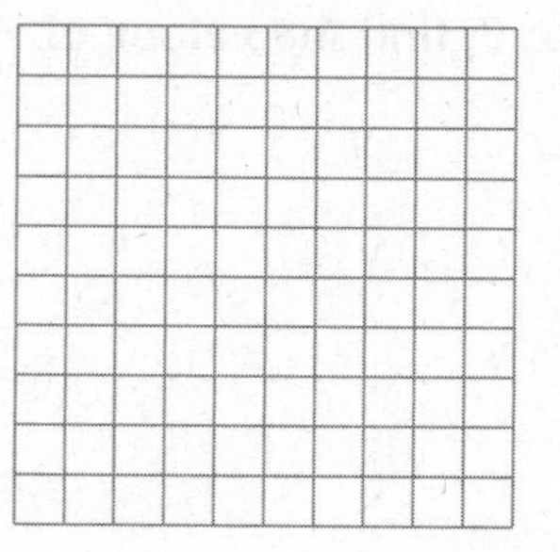

Name________________________________ Date__________

3.3 Completing the Square

For use with Exploration 3.3

Essential Question How can you complete the square for a quadratic expression?

1 EXPLORATION: Using Algebra Tiles to Complete the Square

Go to *BigIdeasMath.com* for an interactive tool to investigate this exploration.

Work with a partner. Use algebra tiles to complete the square for the expression $x^2 + 6x$.

a. You can model $x^2 + 6x$ using one x^2-tile and six x-tiles. Arrange the tiles in a square. Your arrangement will be incomplete in one of the corners.

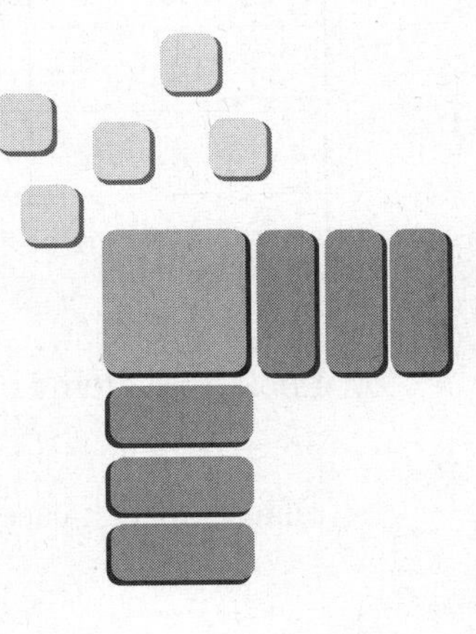

b. How many 1-tiles do you need to complete the square?

c. Find the value of c so that the expression

$x^2 + 6x + c$

is a perfect square trinomial.

d. Write the expression in part (c) as the square of a binomial.

Name ____ Date __________

3.3 Completing the Square (continued)

2 EXPLORATION: Drawing Conclusions

Work with a partner.

a. Use the method outlined in Exploration 1 to complete the table.

Expression	Value of c needed to complete the square	Expression written as a binomial squared
$x^2 + 2x + c$		
$x^2 + 4x + c$		
$x^2 + 8x + c$		
$x^2 + 10x + c$		

b. Look for patterns in the last column of the table. Consider the general statement $x^2 + bx + c = (x + d)^2$. How are d and b related in each case? How are c and d related in each case?

c. How can you obtain the values in the second column directly from the coefficients of x in the first column?

Communicate Your Answer

3. How can you complete the square for a quadratic expression?

4. Describe how you can solve the quadratic equation $x^2 + 6x = 1$ by completing the square.

Name______________________________ Date__________

3.3 Notetaking with Vocabulary

For use after Lesson 3.3

In your own words, write the meaning of each vocabulary term.

completing the square

Core Concepts

Completing the Square

Words To complete the square for the expression $x^2 + bx$, add $\left(\frac{b}{2}\right)^2$.

Diagrams In each diagram, the combined area of the shaded regions is $x^2 + bx$. Adding $\left(\frac{b}{2}\right)^2$ completes the square in the second diagram.

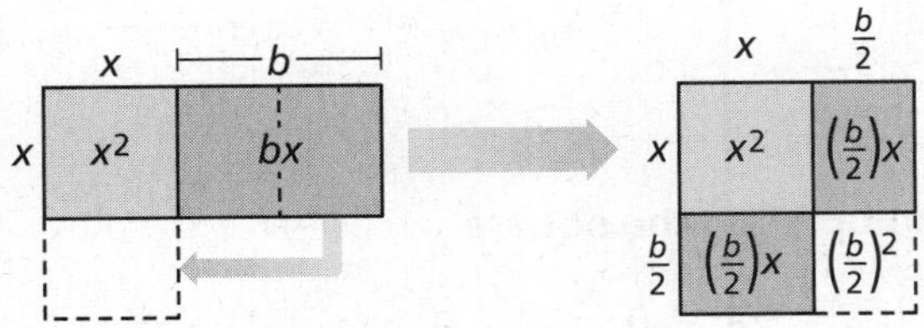

x, b, x, x^2, bx

x, $\frac{b}{2}$, x, x^2, $\left(\frac{b}{2}\right)x$, $\frac{b}{2}$, $\left(\frac{b}{2}\right)x$, $\left(\frac{b}{2}\right)^2$

Algebra $x^2 + bx + \left(\frac{b}{2}\right)^2 = \left(x + \frac{b}{2}\right)\left(x + \frac{b}{2}\right) = \left(x + \frac{b}{2}\right)^2$

Notes:

Name ______________________________ Date __________

Extra Practice

In Exercises 1–3, solve the equation using square roots. Check your solution(s).

1. $x^2 + 4x + 4 = 2$
2. $t^2 - 40t + 400 = 300$
3. $9w^2 + 6w + 1 = -18$

In Exercises 4–6, find the value of *c* that makes the expression a perfect square trinomial. Then write the expression as the square of a binomial.

4. $y^2 - 14y + c$
5. $s^2 + 17s + c$
6. $z^2 + 24z + c$

In Exercises 7–12, solve the equation by completing the square.

7. $r^2 - 6r - 2 = 0$
8. $x^2 + 10x + 28 = 0$
9. $y(y + 1) = \frac{3}{4}$
10. $2t^2 + 16t - 6 = 0$
11. $3x(2x + 10) = -24$
12. $4x^2 - 5x + 28 = 3x^2 + x$

13. Explain how the expression $(4p + 1)^2 + 8(4p + 1) + 16$ is a perfect square trinomial. Then write the expression as a square of a binomial.

In Exercises 14–17, determine whether you would use factoring, square roots, or completing the square to solve the equation. Explain your reasoning. Then solve the equation.

14. $x^2 + 7x = 0$ **15.** $(x - 1)^2 = 35$ **16.** $x^2 - 225 = 0$ **17.** $4x^2 + 8x + 12 = 0$

18. The area of the triangle is 30. Find the value of x.

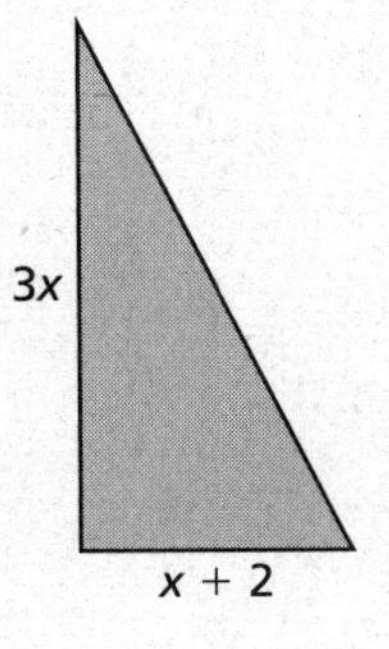

19. Write the quadratic function $f(x) = x^2 + 6x + 22$ in vertex form. Then identify the vertex.

20. A golfer hits a golf ball on the fairway with an initial velocity of 80 feet per second. The height h (in feet) of the golf ball t seconds after it is hit can be modeled by the function $h(t) = -16t^2 + 80t + 0.1$.

a. Find the maximum height of the golf ball.

b. How long does the ball take to hit the ground?

Name ____ Date __________

3.4 Using the Quadratic Formula

For use with Exploration 3.4

Essential Question How can you derive a general formula for solving a quadratic equation?

1 EXPLORATION: Deriving the Quadratic Formula

Work with a partner. Analyze and describe what is done in each step in the development of the Quadratic Formula.

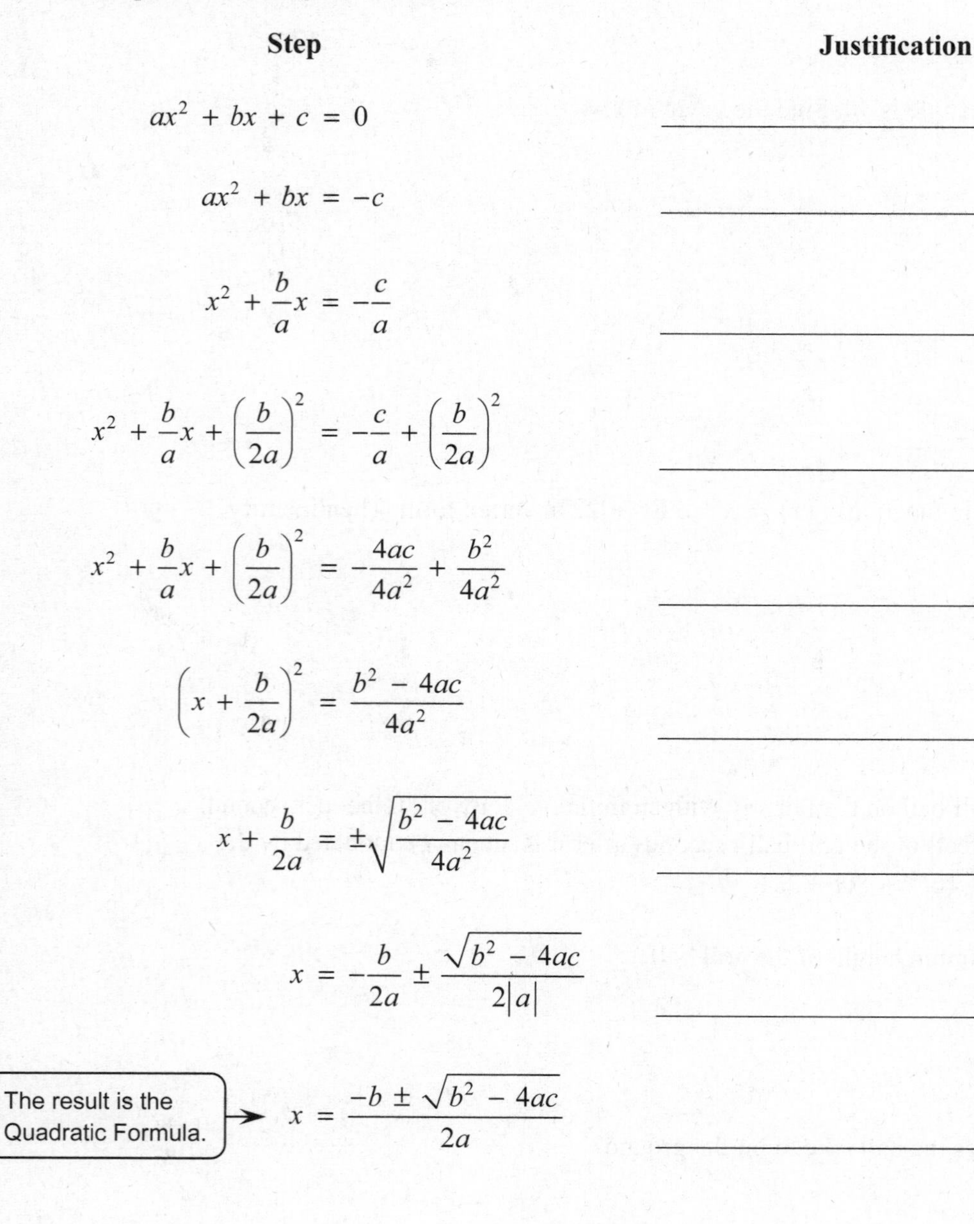

Step	Justification
$ax^2 + bx + c = 0$	____________________
$ax^2 + bx = -c$	____________________
$x^2 + \frac{b}{a}x = -\frac{c}{a}$	____________________
$x^2 + \frac{b}{a}x + \left(\frac{b}{2a}\right)^2 = -\frac{c}{a} + \left(\frac{b}{2a}\right)^2$	____________________
$x^2 + \frac{b}{a}x + \left(\frac{b}{2a}\right)^2 = -\frac{4ac}{4a^2} + \frac{b^2}{4a^2}$	____________________
$\left(x + \frac{b}{2a}\right)^2 = \frac{b^2 - 4ac}{4a^2}$	____________________
$x + \frac{b}{2a} = \pm\sqrt{\frac{b^2 - 4ac}{4a^2}}$	____________________
$x = -\frac{b}{2a} \pm \frac{\sqrt{b^2 - 4ac}}{2\lvert a \rvert}$	____________________
The result is the Quadratic Formula. → $x = \frac{-b \pm \sqrt{b^2 - 4ac}}{2a}$	____________________

Name____ Date__________

3.4 Using the Quadratic Formula (continued)

2 EXPLORATION: Using the Quadratic Formula

Work with a partner. Use the Quadratic Formula to solve each equation.

a. $x^2 - 4x + 3 = 0$

b. $x^2 - 2x + 2 = 0$

c. $x^2 + 2x - 3 = 0$

d. $x^2 + 4x + 4 = 0$

e. $x^2 - 6x + 10 = 0$

f. $x^2 + 4x + 6 = 0$

Communicate Your Answer

3. How can you derive a general formula for solving a quadratic equation?

4. Summarize the following methods you have learned for solving quadratic equations: graphing, using square roots, factoring, completing the square, and using the Quadratic Formula.

Name ______________________________ Date __________

Notetaking with Vocabulary
For use after Lesson 3.4

In your own words, write the meaning of each vocabulary term.

Quadratic Formula

discriminant

Core Concepts

The Quadratic Formula

Let a, b, and c be real numbers such that $a \neq 0$. The solutions of the quadratic equation $ax^2 + bx + c = 0$ are $x = \dfrac{-b \pm \sqrt{b^2 - 4ac}}{2a}$.

Notes:

Name____ Date_______________

3.4 Notetaking with Vocabulary (continued)

Analyzing the Discriminant of $ax^2 + bx + c = 0$

Value of discriminant	$b^2 - 4ac > 0$	$b^2 - 4ac = 0$	$b^2 - 4ac < 0$
Number and type of solutions	Two real solutions	One real solution	Two imaginary solutions
Graph of $y = ax^2 + bx + c$	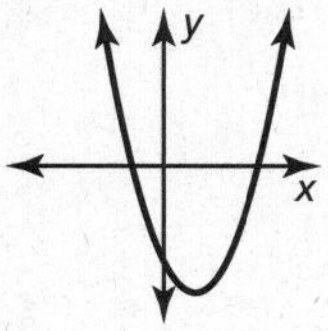 Two x-intercepts	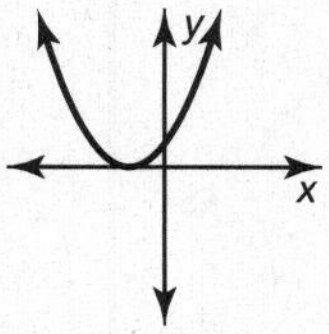 One x-intercept	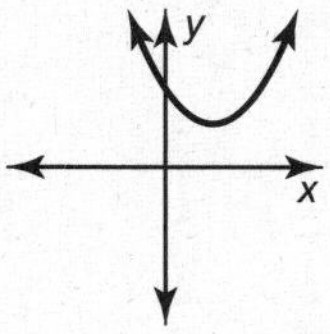 No x-intercept

Notes:

Extra Practice

In Exercises 1–3, solve the equation using the Quadratic Formula. Use a graphing calculator to check your solution(s).

1. $x^2 - 7x - 18 = 0$

2. $w^2 = 4w - 1$

3. $-7z = -4z^2 - 3$

In Exercises 4–6, find the discriminant of the quadratic equation and describe the number and type of solutions of the equation.

4. $b^2 + 34b + 289 = 0$

5. $x^2 = 3 - 8x$

6. $4q^2 + 1 = 3q$

7. A baseball player hits a foul ball straight up in the air from a height of 4 feet off the ground with an initial velocity of 85 feet per second.

a. Write a quadratic function that represents the height h of the ball t seconds after it hits the bat.

b. When is the ball 110 feet off the ground? Explain your reasoning.

c. The catcher catches the ball 6 feet from the ground. How long is the ball in the air?

Name____ Date__________

3.5 Solving Nonlinear Systems

For use with Exploration 3.5

Essential Question How can you solve a nonlinear system of equations?

EXPLORATION: Solving Nonlinear Systems of Equations

Work with a partner. Match each system with its graph. Explain your reasoning. Then solve each system using the graph.

a. $y = x^2$
$y = x + 2$

b. $y = x^2 + x - 2$
$y = x + 2$

c. $y = x^2 - 2x - 5$
$y = -x + 1$

d. $y = x^2 + x - 6$
$y = -x^2 - x + 6$

e. $y = x^2 - 2x + 1$
$y = -x^2 + 2x - 1$

f. $y = x^2 + 2x + 1$
$y = -x^2 + x + 2$

A.

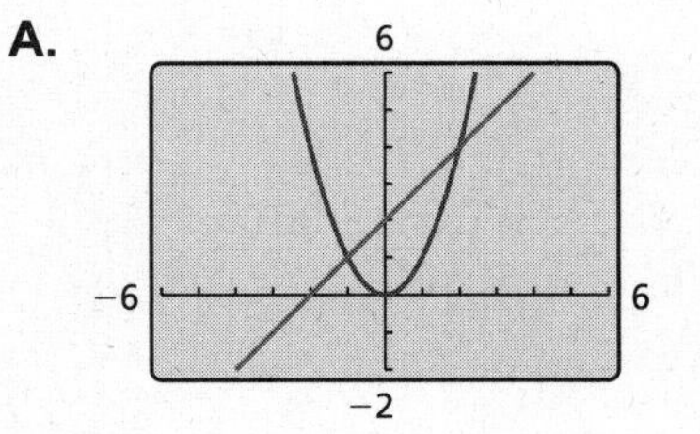

B.

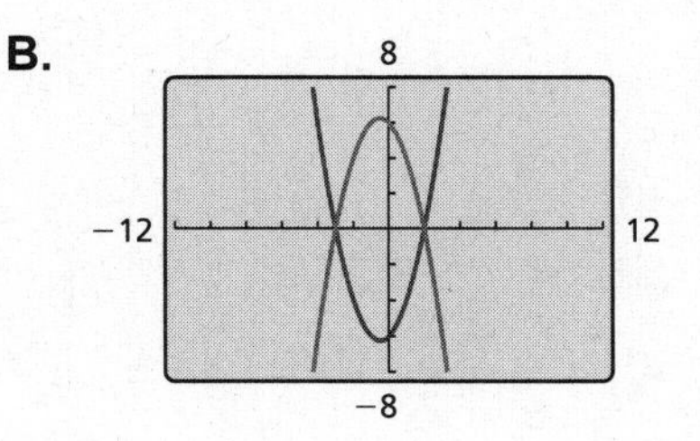

C.

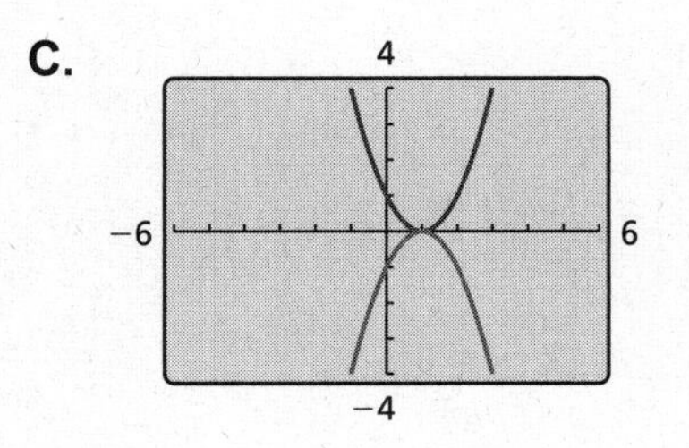

D.

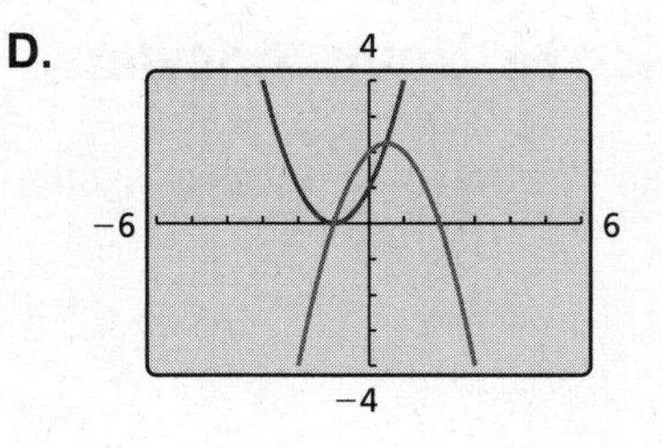

E.

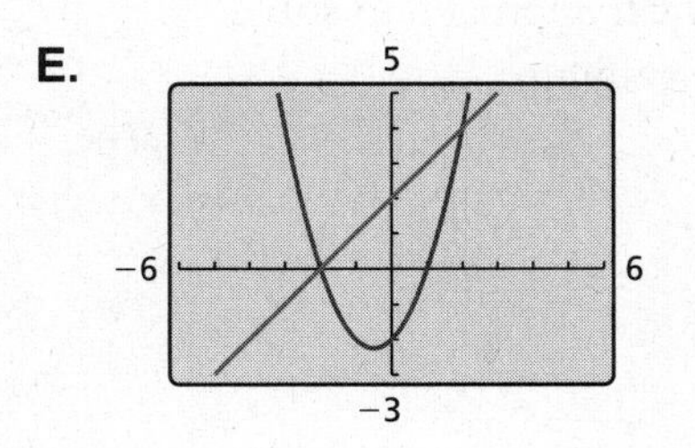

F.

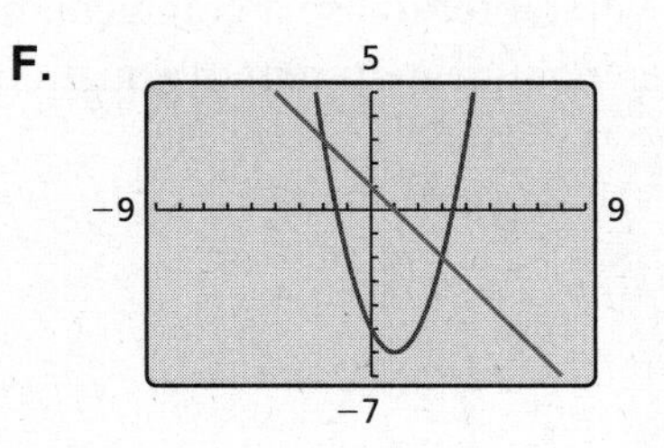

2 EXPLORATION: Solving Nonlinear Systems of Equations

Go to *BigIdeasMath.com* for an interactive tool to investigate this exploration.

Work with a partner. Look back at the nonlinear system in Exploration 1(f). Suppose you want a more accurate way to solve the system than using a graphical approach.

a. Show how you could use a *numerical approach* by creating a table. For instance, you might use a spreadsheet to solve the system.

b. Show how you could use an *analytical approach*. For instance, you might try solving the system by substitution or elimination.

Communicate Your Answer

3. How can you solve a nonlinear system of equations?

4. Would you prefer to use a graphical, numerical, or analytical approach to solve the given nonlinear system of equations? Explain your reasoning.

$$y = x^2 + 2x - 3$$
$$y = -x^2 - 2x + 4$$

Name____ Date__________

3.5 Notetaking with Vocabulary
For use after Lesson 3.5

In your own words, write the meaning of each vocabulary term.

system of nonlinear equations

Core Concepts

Solve Equations by Graphing

Step 1 To solve the equation $f(x) = g(x)$, write a system of two equations, $y = f(x)$ and $y = g(x)$.

Step 2 Graph the system of equations $y = f(x)$ and $y = g(x)$. The x-value of each solution of the system is a solution of the equation $f(x) = g(x)$.

Notes:

3.5 Notetaking with Vocabulary (continued)

Extra Practice

In Exercises 1–3, solve the system by graphing. Check your solution(s).

1. $y = \frac{1}{2}x^2 - 3$
$y = -4 - 2x^2$

2. $y = (x - 2)^2$
$y = \frac{1}{4}x - \frac{1}{2}$

3. $y = -x^2 - 2$
$y = 4(x + 1) - 3$

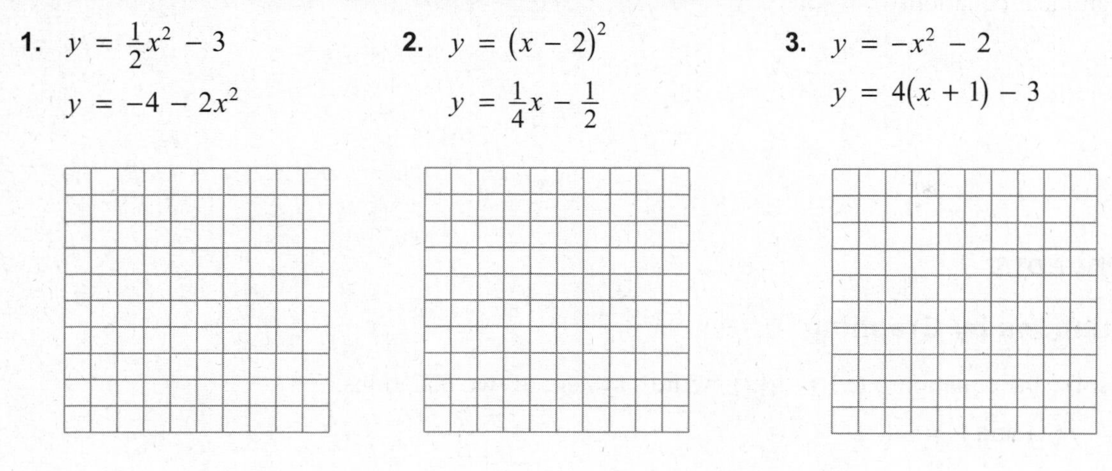

In Exercises 4 and 5, solve the system of nonlinear equations by using the graph.

4.

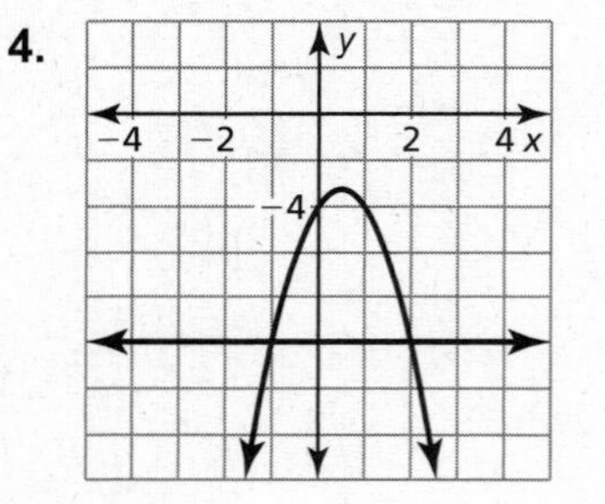

5.

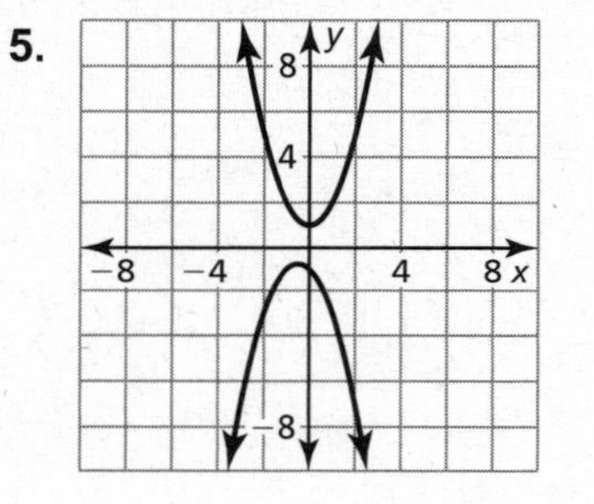

In Exercises 6–8, solve the system by substitution.

6. $y = x + 4$
$y = (x + 2)^2 + 1$

7. $x^2 + y^2 = 16$
$y = -x + 4$

8. $2x^2 + 10x + 48 = y - 10x$
$-4x^2 - 16x = y$

3.5 Notetaking with Vocabulary (continued)

In Exercises 9–11, solve the system by elimination.

9. $x^2 - 7x + 11 = y - 1$
$-x + y = -4$

10. $y = 9x^2 + 6x + 2$
$y = x^2 - 8x - 19$

11. $-5x + 29 = y - x^2$
$x^2 + y = 2x^2 - 1$

12. Consider the following system.

$x^2 = 9 - y^2$
$x + 2y = 2x^2 + 7 + x$

a. Which method would you use to solve the system? Explain your reasoning.

b. Would you have used a different method if the system had been as follows? Explain.

$x = 9 - y$
$x + 2y = 2x^2 + 7 + x$

13. The sum of two numbers is -5, and the sum of the squares of the two numbers is 17. What are the two numbers? Explain your reasoning.

Name ______________________________ Date __________

3.6 Quadratic Inequalities

For use with Exploration 3.6

Essential Question How can you solve a quadratic inequality?

1 EXPLORATION: Solving a Quadratic Inequality

Work with a partner. The graphing calculator screen shows the graph of

$f(x) = x^2 + 2x - 3.$

Explain how you can use the graph to solve the inequality

$x^2 + 2x - 3 \leq 0.$

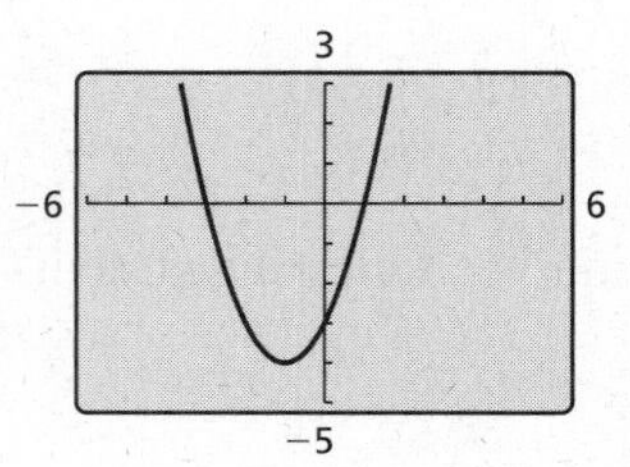

Then solve the inequality.

2 EXPLORATION: Solving Quadratic Inequalities

Work with a partner. Match each inequality with the graph of its related quadratic function on the next page. Then use the graph to solve the inequality.

a. $x^2 - 3x + 2 > 0$ **b.** $x^2 - 4x + 3 \leq 0$ **c.** $x^2 - 2x - 3 < 0$

d. $x^2 + x - 2 \geq 0$ **e.** $x^2 - x - 2 < 0$ **f.** $x^2 - 4 > 0$

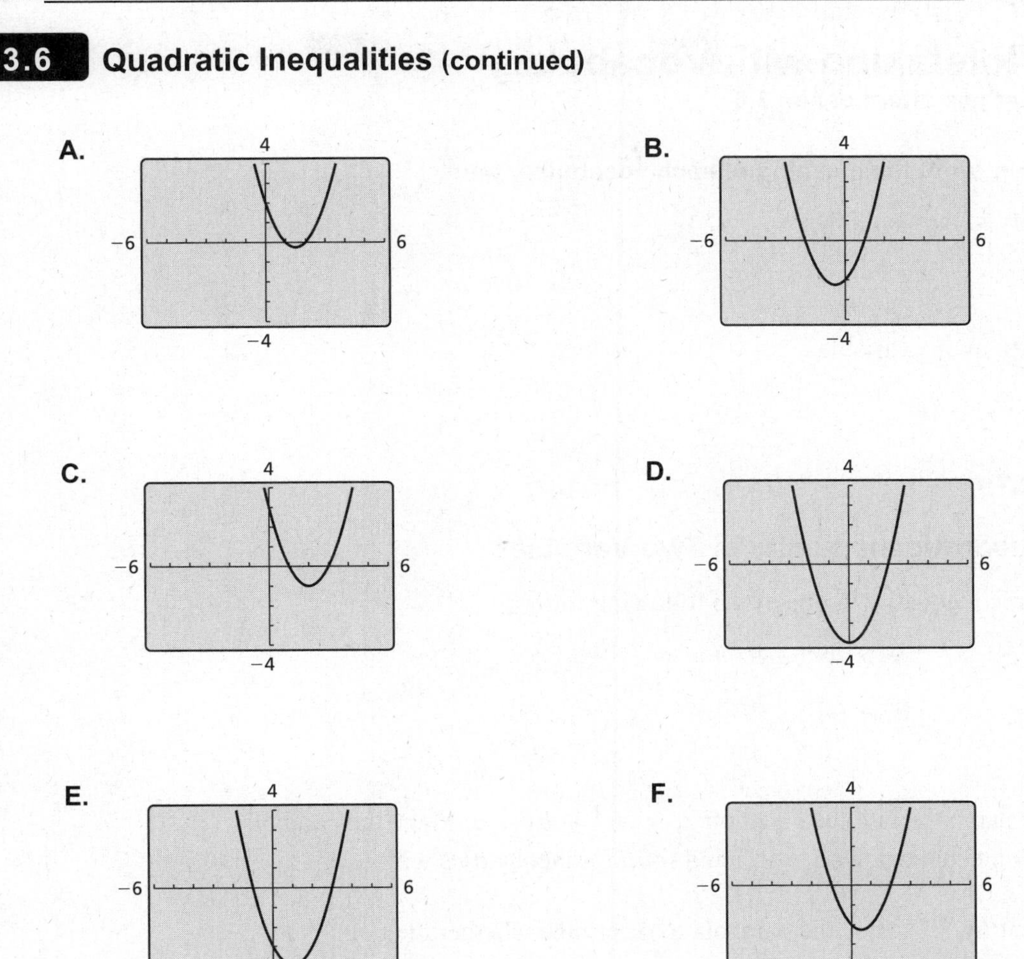

Communicate Your Answer

3. How can you solve a quadratic inequality?

4. Explain how you can use the graph in Exploration 1 to solve each inequality. Then solve each inequality.

a. $x^2 + 2x - 3 > 0$ **b.** $x^2 + 2x - 3 < 0$ **c.** $x^2 + 2x - 3 \geq 0$

Name ______________________________ Date __________

3.6 Notetaking with Vocabulary

For use after Lesson 3.6

In your own words, write the meaning of each vocabulary term.

quadratic inequality in two variables

quadratic inequality in one variable

Core Concepts

Graphing a Quadratic Inequality in Two Variables

To graph a quadratic inequality in one of the following forms,

$y < ax^2 + bx + c$ $\qquad$ $y > ax^2 + bx + c$

$y \le ax^2 + bx + c$ $\qquad$ $y \ge ax^2 + bx + c,$

follow these steps.

Step 1 Graph the parabola with the equation $y = ax^2 + bx + c$. Make the parabola *dashed* for inequalities with $<$ or $>$ and *solid* for inequalities with $\le$ or $\ge$.

Step 2 Test a point (x, y) inside the parabola to determine whether the point is a solution of the inequality.

Step 3 Shade the region inside the parabola if the point from Step 2 is a solution. Shade the region outside the parabola if it is not a solution.

Notes:

Name____ Date__________

Extra Practice

In Exercises 1–4, match the graph with its inequality. Explain your reasoning.

1. 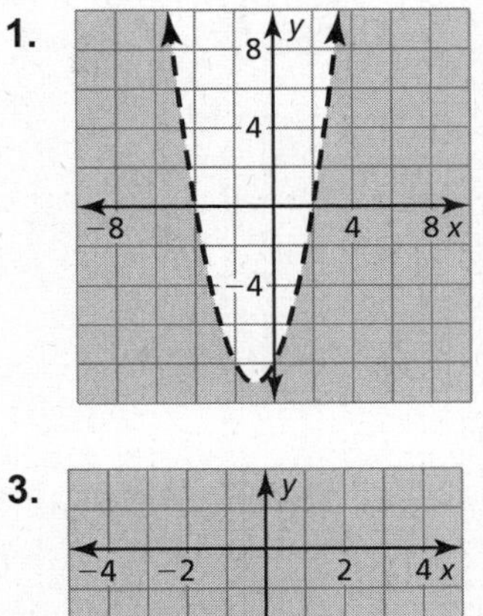

2. 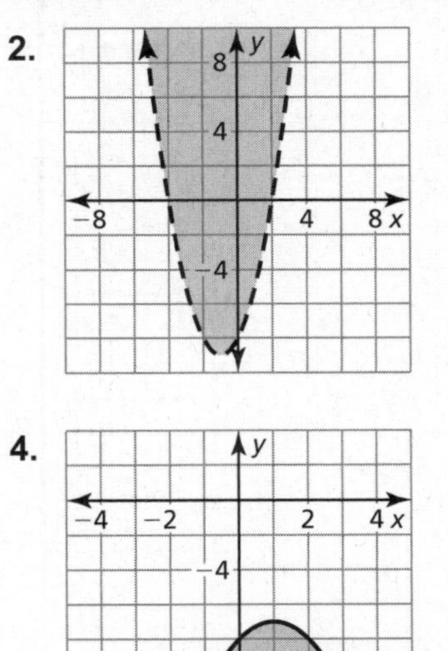

3. 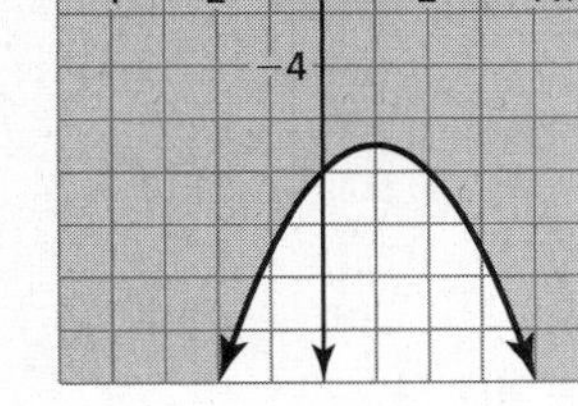

4. 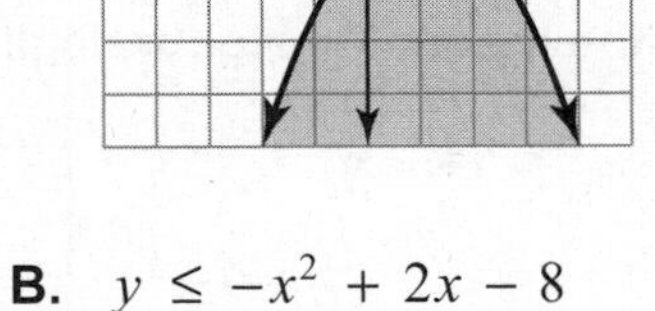

A. $y < x^2 + 2x - 8$

B. $y \leq -x^2 + 2x - 8$

C. $y > x^2 + 2x - 8$

D. $y \geq -x^2 + 2x - 8$

In Exercises 5–8, graph the inequality.

5. $y < x^2 + 2$

6. $y \leq -5x^2$

7. $y \geq -(x + 4)^2 - 1$

8. $y < 4x^2 + 4x + 1$

9. Accident investigators use the formula $d = 0.01875v^2$, where d is the braking distance of a car (in feet) and v is the speed of the car (in miles per hour) to determine how fast a car is going at the time of an accident. For what speeds v would a car leave a tire mark on the road of over 1 foot?

Name ______________________________ Date __________

3.6 Notetaking with Vocabulary (continued)

In Exercises 10–12, graph the system of quadratic inequalities.

10. $y \leq -x^2$
$y > -3x^2 + 3$

11. $y \geq x^2 + 5x$
$y \geq (x + 2)^2 - 1$

12. $y > x^2 - 7x - 8$
$y < -x^2 + 6x + 5$

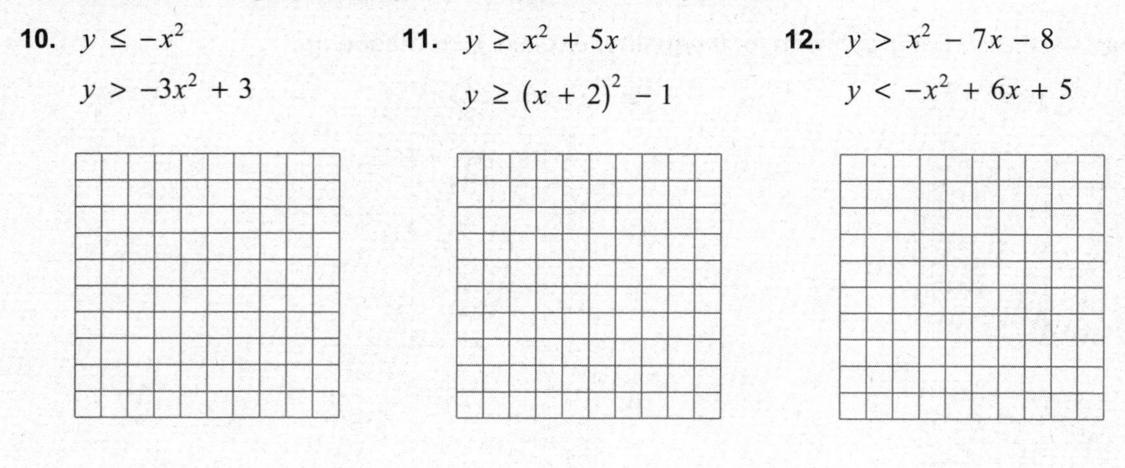

In Exercises 13–15, solve the inequality algebraically.

13. $16x^2 > 100$

14. $x^2 \leq 15x - 34$

15. $-\frac{1}{5}x^2 + 10x \geq -25$

16. The profit for a hot dog company is given by the equation $y = -0.02x^2 + 140x - 2500$, where x is the number of hot dogs produced and y is the profit (in dollars). How many hot dogs must be produced so that the company will generate a profit of at least \$150,000?

Name____________________ Date__________

Chapter 4 Maintaining Mathematical Proficiency

Simplify the expression.

1. $-8x - 9x$

2. $25r - 5 + 7r - r$

3. $3 + 6(3x - 5) + x$

4. $3y - (2y - 5) + 11$

5. $-3(h + 7) - 7(10 - h)$

6. $5 - 8x^2 + 5x + 8x^2$

Find the volume or surface area of the solid.

7. volume of a right cylinder with radius 5 feet and height 15 feet

8. surface area of a rectangular prism with length 10 meters, width 20 meters, and height 4 meters

9. volume of a cube with side length 2.5 millimeters

10. surface area of a sphere with radius 1 foot

11. For what radius length can the value of the volume of a sphere equal the value of the surface area?

Name ______________________________ Date __________

4.1 Graphing Polynomial Functions

For use with Exploration 4.1

Essential Question What are some common characteristics of the graphs of cubic and quartic polynomial functions?

1 EXPLORATION: Identifying Graphs of Polynomial Functions

Go to *BigIdeasMath.com* for an interactive tool to investigate this exploration.

Work with a partner. Match each polynomial function with its graph. Explain your reasoning. Use a graphing calculator to verify your answers.

a. $f(x) = x^3 - x$ **b.** $f(x) = -x^3 + x$ **c.** $f(x) = -x^4 + 1$

d. $f(x) = x^4$ **e.** $f(x) = x^3$ **f.** $f(x) = x^4 - x^2$

A.

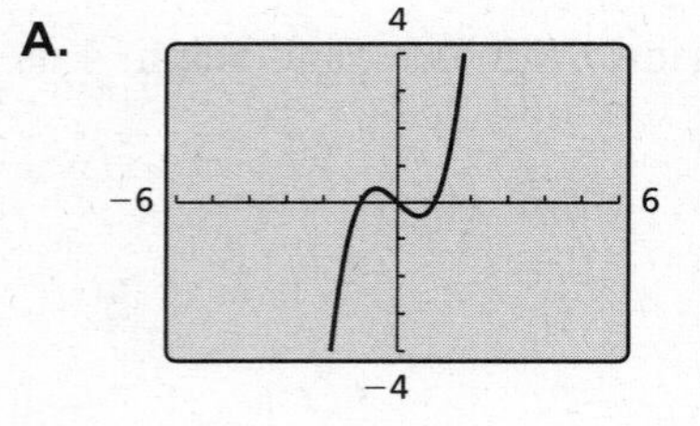

B.

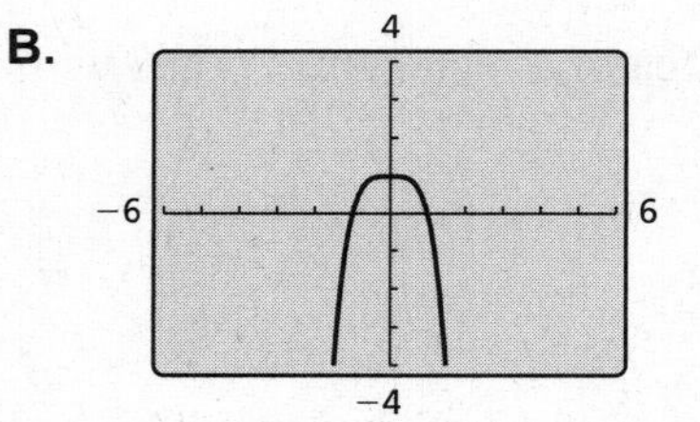

C.

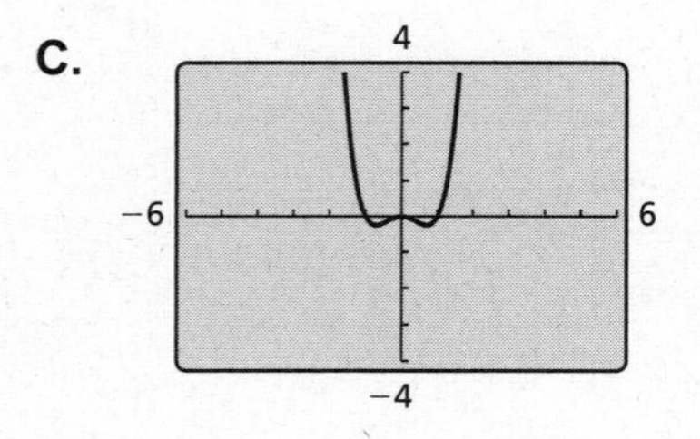

D.

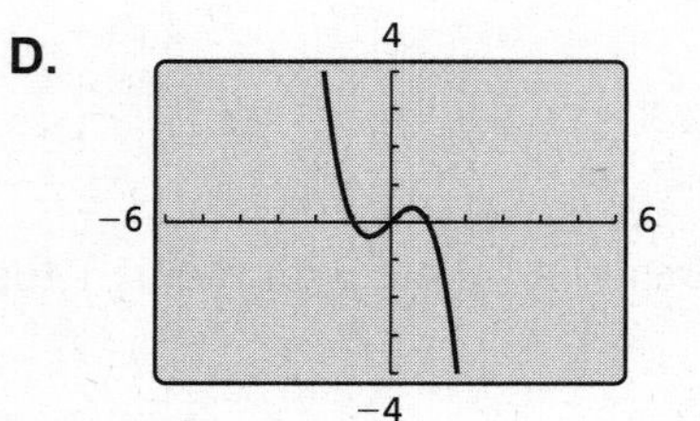

E.

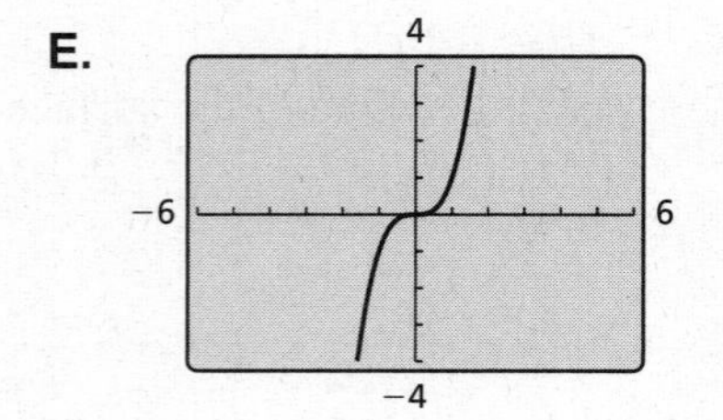

F.

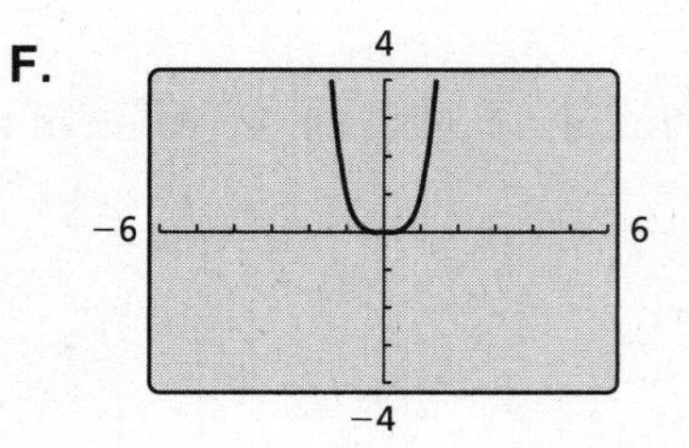

Name____ Date__________

2 EXPLORATION: Identifying x-Intercepts of Polynomial Graphs

Work with a partner. Each of the polynomial graphs in Exploration 1 has x-intercept(s) of -1 , 0, or 1. Identify the x-intercept(s) of each graph. Explain how you can verify your answers.

Communicate Your Answer

3. What are some common characteristics of the graphs of cubic and quartic polynomial functions?

4. Determine whether each statement is *true* or *false*. Justify your answer.

a. When the graph of a cubic polynomial function rises to the left, it falls to the right.

b. When the graph of a quartic polynomial function falls to the left, it rises to the right.

Name ____ Date __________

4.1 Notetaking with Vocabulary
For use after Lesson 4.1

In your own words, write the meaning of each vocabulary term.

polynomial

polynomial function

end behavior

Core Concepts

End Behavior of Polynomial Functions

Degree: odd

Leading coefficient: positive

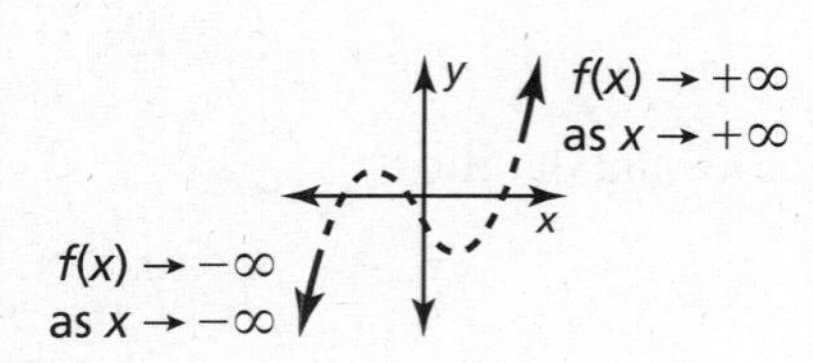

Degree: odd

Leading coefficient: negative

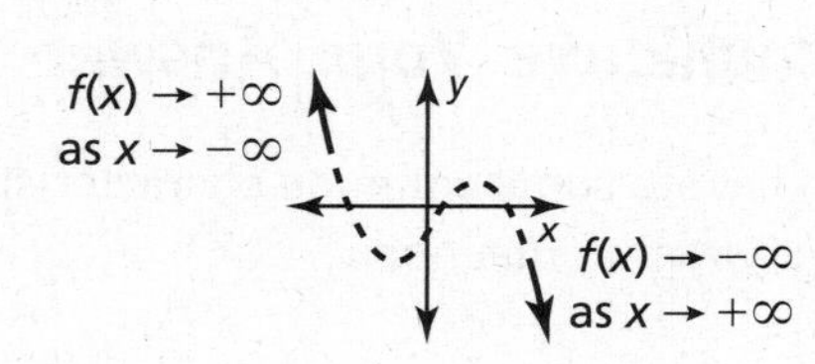

Degree: even

Leading coefficient: positive

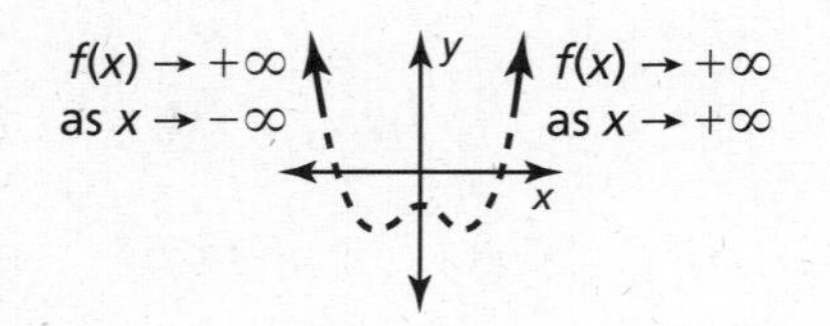

Degree: even

Leading coefficient: negative

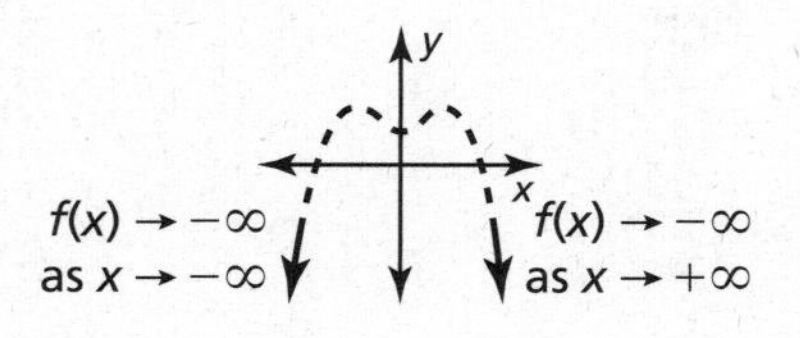

Notes:

Name____ Date__________

Extra Practice

In Exercises 1–4, decide whether the function is a polynomial function. If so, write it in standard form and state its degree, type, and leading coefficient.

1. $f(x) = 2x^2 - 3x^4 + 6x + 1$

2. $m(x) = -\frac{3}{7}x^3 + \frac{7}{x} - 3$

3. $g(x) = \sqrt{15}x + \sqrt{5}$

4. $p(x) = -2\sqrt{3} + 3x - 2x^2$

In Exercises 5 and 6, evaluate the function for the given value of *x*.

5. $h(x) = -x^3 - 2x^2 - 3x + 4;\ x = 2$

6. $g(x) = x^4 - 32x^2 + 256;\ x = -4$

In Exercises 7 and 8, describe the end behavior of the graph of the function.

7. $f(x) = -3x^6 + 4x^2 - 3x + 6$

8. $f(x) = \frac{4}{5}x + 6x + 3x^5 - 3x^3 - 2$

9. Describe the degree and leading coefficient of the polynomial function using the graph.

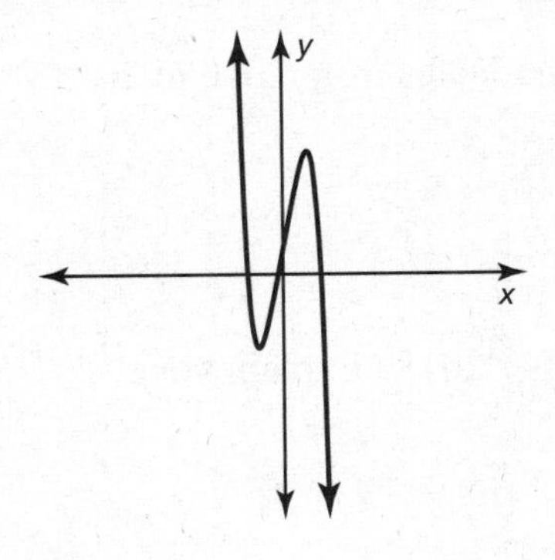

4.1 Notetaking with Vocabulary (continued)

In Exercises 10 and 11, graph the polynomial function.

10. $p(x) = 16 - x^4$

11. $g(x) = x^2 + 3x^5 - x$

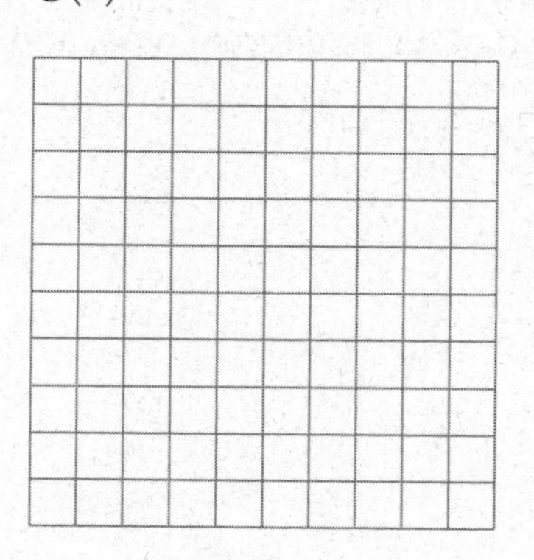

12. Sketch a graph of the polynomial function f if

f is increasing when $x < -1$ and $0 < x < 1$,

f is decreasing when $-1 < x < 0$ and $x > 1$,

and $f(x) < 0$ for all real numbers.

Describe the degree and leading coefficient of the function f.

13. The number of students S (in thousands) who graduate in four years from a university can be modeled by the function $S(t) = -\frac{1}{4}t^3 + t^2 + 23$, where t is the number of years since 2010.

a. Use a graphing calculator to graph the function for the interval $0 \leq t \leq 5$. Describe the behavior of the graph on this interval.

b. What is the average rate of change in the number of four-year graduates from 2010 to 2015?

c. Do you think this model can be used for years before 2010 or after 2015? Explain your reasoning.

Name____ Date__________

4.2 Adding, Subtracting, and Multiplying Polynomials

For use with Exploration 4.2

Essential Question How can you cube a binomial?

1 EXPLORATION: Cubing Binomials

Work with a partner. Find each product. Show your steps.

a. $(x + 1)^3 = (x + 1)(x + 1)^2$ Rewrite as a product of first and second powers.

$= (x + 1)$ __________ Multiply second power.

$=$ __________ Multiply binomial and trinomial.

$=$ __________ Write in standard form, $ax^3 + bx^2 + cx + d$.

b. $(a + b)^3 = (a + b)(a + b)^2$ Rewrite as a product of first and second powers.

$= (a + b)$ __________ Multiply second power.

$=$ __________ Multiply binomial and trinomial.

$=$ __________ Write in standard form.

c. $(x - 1)^3 = (x - 1)(x - 1)^2$ Rewrite as a product of first and second powers.

$= (x - 1)$ __________ Multiply second power.

$=$ __________ Multiply binomial and trinomial.

$=$ __________ Write in standard form.

d. $(a - b)^3 = (a - b)(a - b)^2$ Rewrite as a product of first and second powers.

$= (a - b)$ __________ Multiply second power.

$=$ __________ Multiply binomial and trinomial.

$=$ __________ Write in standard form.

2 EXPLORATION: Generalizing Patterns for Cubing a Binomial

Work with a partner.

a. Use the results of Exploration 1 to describe a pattern for the coefficients of the terms when you expand the cube of a binomial. How is your pattern related to Pascal's Triangle, shown at the right?

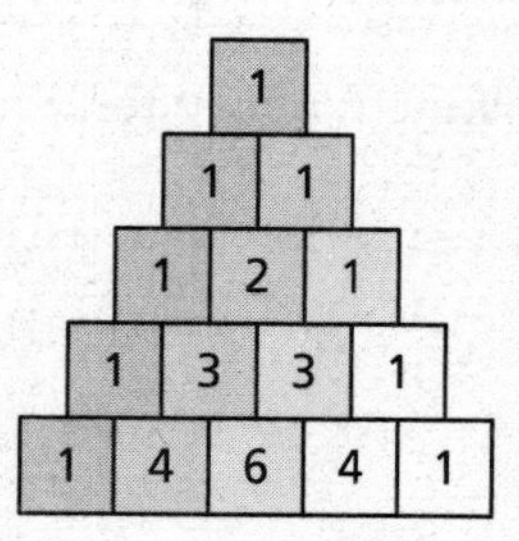

b. Use the results of Exploration 1 to describe a pattern for the exponents of the terms in the expansion of a cube of a binomial.

c. Explain how you can use the patterns you described in parts (a) and (b) to find the product $(2x - 3)^3$. Then find this product.

Communicate Your Answer

3. How can you cube a binomial?

4. Find each product.

a. $(x + 2)^3$ **b.** $(x - 2)^3$ **c.** $(2x - 3)^3$

d. $(x - 3)^3$ **e.** $(-2x + 3)^3$ **f.** $(3x - 5)^3$

Name____ Date __________

4.2 Notetaking with Vocabulary
For use after Lesson 4.2

In your own words, write the meaning of each vocabulary term.

Pascal's Triangle

Core Concepts

Special Product Patterns

Sum and Difference

$(a + b)(a - b) = a^2 - b^2$

Example

$(x + 3)(x - 3) = x^2 - 9$

Square of a Binomial

$(a + b)^2 = a^2 + 2ab + b^2$

$(a - b)^2 = a^2 - 2ab + b^2$

Example

$(y + 4)^2 = y^2 + 8y + 16$

$(2t - 5)^2 = 4t^2 - 20t + 25$

Cube of a Binomial

$(a + b)^3 = a^3 + 3a^2b + 3ab^2 + b^3$

$(a - b)^3 = a^3 - 3a^2b + 3ab^2 - b^3$

Example

$(z + 3)^3 = z^3 + 9z^2 + 27z + 27$

$(m - 2)^3 = m^3 - 6m^2 + 12m - 8$

Notes:

Name ______________________________ Date __________

4.2 Notetaking with Vocabulary (continued)

Pascal's Triangle

In Pascal's Triangle, the first and last numbers in each row are 1. Every number other than 1 is the sum of the closest two numbers in the row directly above it. The numbers in Pascal's Triangle are the same numbers that are the coefficients of binomial expansions, as shown in the first six rows.

	n	$(a+b)^n$	Binomial Expansion	Pascal's Triangle
0th row	0	$(a+b)^0 =$	1	1
1st row	1	$(a+b)^1 =$	$1a + 1b$	1 1
2nd row	2	$(a+b)^2 =$	$1a^2 + 2ab + 1b^2$	1 2 1
3rd row	3	$(a+b)^3 =$	$1a^3 + 3a^2b + 3ab^2 + 1b^3$	1 3 3 1
4th row	4	$(a+b)^4 =$	$1a^4 + 4a^3b + 6a^2b^2 + 4ab^3 + 1b^4$	1 4 6 4 1
5th row	5	$(a+b)^5 =$	$1a^5 + 5a^4b + 10a^3b^2 + 10a^2b^3 + 5ab^4 + 1b^5$	1 5 10 10 5 1

Notes:

Name______________________________ Date__________

Extra Practice

In Exercises 1–3, find the sum or difference.

1. $(-4x^2 - 6x + 18) + (-x^2 + 7x + 8)$

2. $(6x^2 - 12x + 48) - (-x^2 + 24x - 63)$

3. $(-11x^4 - x^3 - 3x^2 + 10x - 2) - (-11x^4 + 5x^2 - 7x + 13)$

In Exercises 4–9, find the product.

4. $2x^2(2x^3 - x^2 + 3x - 5)$

5. $(x^4 - 10x^2 + 25)(3x^2 - 6x - 1)$

6. $(x + 1)(x - 2)(x + 6)$

7. $(2x - 3)(6 - x)(4 - 5x)$

8. $(3y - 8)(3y + 8)$

9. $(2v - 1)^3$

In Exercises 10 and 11, use Pascal's Triangle to expand the binomial.

10. $(4t - 2)^4$

11. $(g + 6)^5$

Name ______________________________ Date __________

4.3 Dividing Polynomials

For use with Exploration 4.3

Essential Question How can you use the factors of a cubic polynomial to solve a division problem involving the polynomial?

1 EXPLORATION: Dividing Polynomials

Go to *BigIdeasMath.com* for an interactive tool to investigate this exploration.

Work with a partner. Match each division statement with the graph of the related cubic polynomial $f(x)$. Explain your reasoning. Use a graphing calculator to verify your answers.

a. $\dfrac{f(x)}{x} = (x - 1)(x + 2)$

b. $\dfrac{f(x)}{x - 1} = (x - 1)(x + 2)$

c. $\dfrac{f(x)}{x + 1} = (x - 1)(x + 2)$

d. $\dfrac{f(x)}{x - 2} = (x - 1)(x + 2)$

e. $\dfrac{f(x)}{x + 2} = (x - 1)(x + 2)$

f. $\dfrac{f(x)}{x - 3} = (x - 1)(x + 2)$

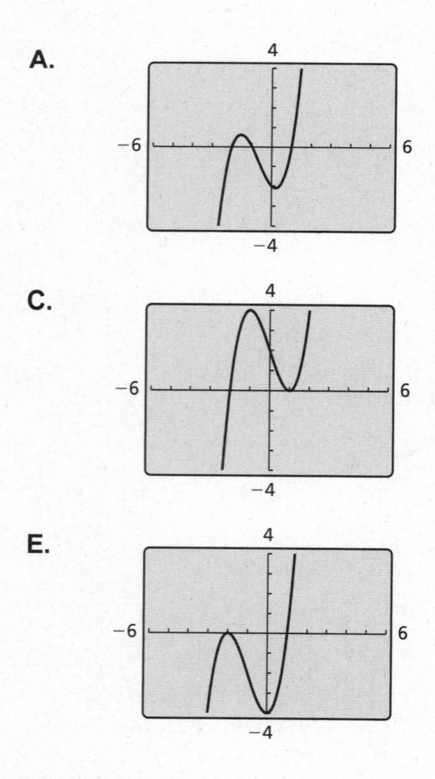

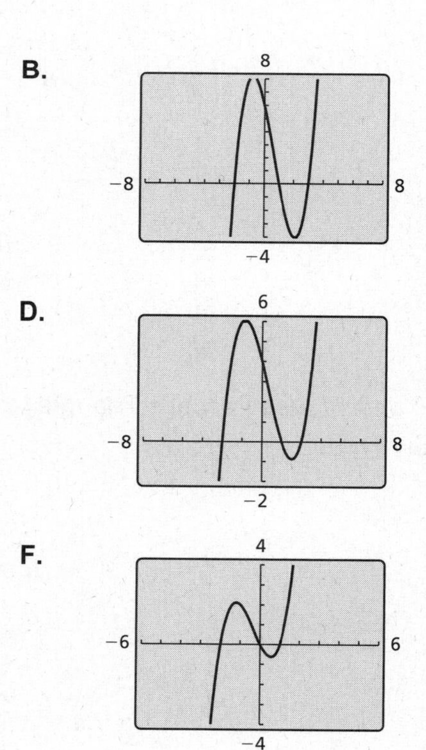

Name______________________________ Date__________

2 EXPLORATION: Dividing Polynomials

Work with a partner. Use the results of Exploration 1 to find each quotient. Write your answers in standard form. Check your answers by multiplying.

a. $(x^3 + x^2 - 2x) \div x$

b. $(x^3 - 3x + 2) \div (x - 1)$

c. $(x^3 + 2x^2 - x - 2) \div (x + 1)$

d. $(x^3 - x^2 - 4x + 4) \div (x - 2)$

e. $(x^3 + 3x^2 - 4) \div (x + 2)$

f. $(x^3 - 2x^2 - 5x + 6) \div (x - 3)$

Communicate Your Answer

3. How can you use the factors of a cubic polynomial to solve a division problem involving the polynomial?

Name ______________________________ Date __________

4.3 Notetaking with Vocabulary
For use after Lesson 4.3

In your own words, write the meaning of each vocabulary term.

polynomial long division

synthetic division

Core Concepts

The Remainder Theorem

If a polynomial $f(x)$ is divided by $x - k$, then the remainder is $r = f(k)$.

Notes:

4.3 Notetaking with Vocabulary (continued)

Extra Practice

In Exercises 1–4, divide using polynomial long division.

1. $(x^2 + 6x + 12) \div (x - 3)$

2. $(x^3 - 4x^2) \div (x^2 - 16)$

3. $(4x^3 + 13x^2 + 27x + 6) \div (4x + 1)$

4. $(x^4 + 2x^3 + 5x^2 + 3x) \div (x^2 - x)$

In Exercises 5–8, divide using synthetic division.

5. $(x^2 - 10x + 2) \div (x - 2)$

6. $(x^3 + 4x^2 + 6x + 4) \div (x + 2)$

7. $(2x^3 - 54) \div (x + 3)$

8. $(2x^4 - 11x^3 + 11x^2 + 4x + 4) \div (x - 4)$

4.3 Notetaking with Vocabulary (continued)

In Exercises 9–12, match the equivalent expressions. Justify your answers.

9. $(x^2 - x - 8) \div (x - 4)$

A. $x + 3 + \dfrac{4}{x - 4}$

10. $(x^2 - x + 8) \div (x - 4)$

B. $x + 5 + \dfrac{12}{x - 4}$

11. $(x^2 + x - 8) \div (x - 4)$

C. $x + 5 + \dfrac{28}{x - 4}$

12. $(x^2 + x + 8) \div (x - 4)$

D. $x + 3 + \dfrac{20}{x - 4}$

In Exercises 13–16, use synthetic division to evaluate the function for the indicated value of *x*.

13. $f(x) = -3x^3 + 4x^2 - 17x - 6; x = 2$

14. $f(x) = -x^4 + x^2 + 4; x = -1$

15. $f(x) = x^3 - 10x^2 + 31x - 30; x = -5$

16. $f(x) = x^3 + 8x + 27; x = 3$

17. What is the value of k such that $(-x^4 + 5x^2 + kx - 8) \div (x - 4)$ has a remainder of 0?

Name____ Date__________

4.4 Factoring Polynomials

For use with Exploration 4.4

Essential Question How can you factor a polynomial?

1 EXPLORATION: Factoring Polynomials

Work with a partner. Match each polynomial equation with the graph of its related polynomial function. Use the x-intercepts of the graph to write each polynomial in factored form. Explain your reasoning.

a. $x^2 + 5x + 4 = 0$

b. $x^3 - 2x^2 - x + 2 = 0$

c. $x^3 + x^2 - 2x = 0$

d. $x^3 - x = 0$

e. $x^4 - 5x^2 + 4 = 0$

f. $x^4 - 2x^3 - x^2 + 2x = 0$

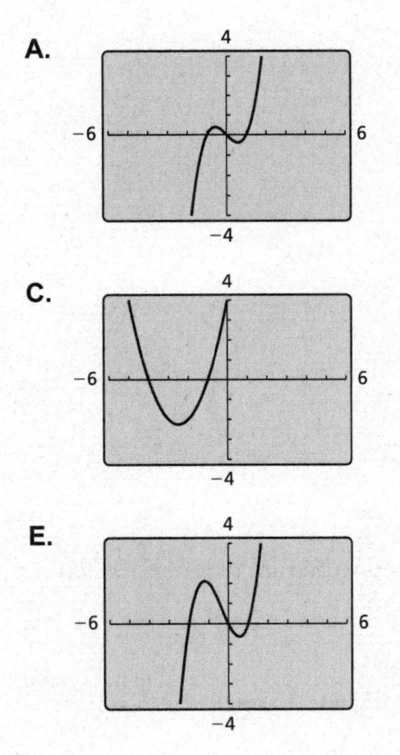

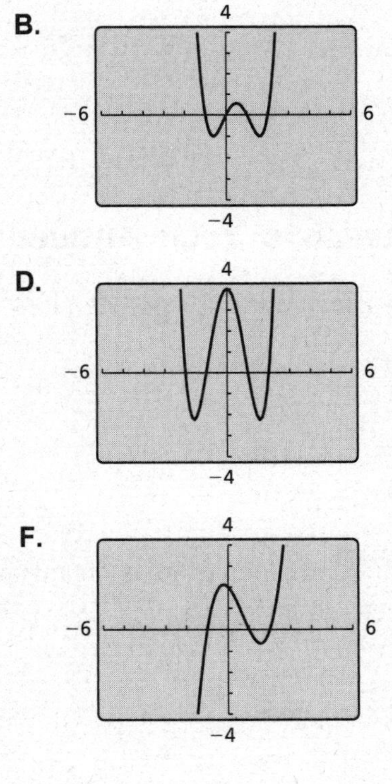

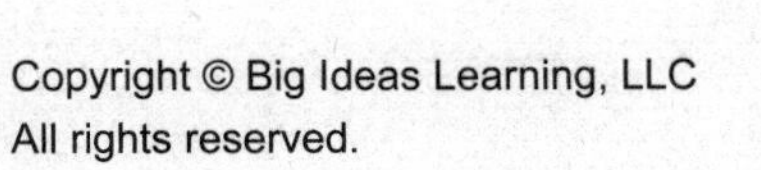

2 EXPLORATION: Factoring Polynomials

Work with a partner. Use the x-intercepts of the graph of the polynomial function to write each polynomial in factored form. Explain your reasoning. Check your answers by multiplying.

a. $f(x) = x^2 - x - 2$

b. $f(x) = x^3 - x^2 - 2x$

c. $f(x) = x^3 - 2x^2 - 3x$

d. $f(x) = x^3 - 3x^2 - x + 3$

e. $f(x) = x^4 + 2x^3 - x^2 - 2x$

f. $f(x) = x^4 - 10x^2 + 9$

Communicate Your Answer

3. How can you factor a polynomial?

4. What information can you obtain about the graph of a polynomial function written in factored form?

Name______________________________ Date__________

4.4 Notetaking with Vocabulary

For use after Lesson 4.4

In your own words, write the meaning of each vocabulary term.

factored completely

factor by grouping

quadratic form

Core Concepts

Special Factoring Patterns

Sum of Two Cubes

$a^3 + b^3 = (a + b)(a^2 - ab + b^2)$

Example

$64x^3 + 1 = (4x)^3 + 1^3$

$= (4x + 1)(16x^2 - 4x + 1)$

Difference of Two Cubes

$a^3 - b^3 = (a - b)(a^2 + ab + b^2)$

Example

$27x^3 - 8 = (3x)^3 - 2^3$

$= (3x - 2)(9x^2 + 6x + 4)$

Notes:

Name ______________________________ Date __________

4.4 Notetaking with Vocabulary (continued)

The Factor Theorem

A polynomial $f(x)$ has a factor $x - k$ if and only if $f(k) = 0$.

Notes:

Extra Practice

In Exercises 1–14, factor the polynomial completely.

1. $20x^3 - 220x^2 + 600x$

2. $m^5 - 81m$

3. $27a^3 + 8b^3$

4. $5t^6 + 2t^5 - 5t^4 - 2t^3$

5. $y^4 - 13y^2 - 48$

6. $5p^3 + 5p - 5p^2 - 5$

7. $810k^4 - 160$

8. $a^5 + a^3 - a^2 - 1$

Name______________________ Date__________

4.4 Notetaking with Vocabulary (continued)

9. $2x^6 - 8x^5 - 42x^4$

10. $5z^3 + 5z^2 - 6z - 6$

11. $12x^2 - 22x - 20$

12. $3m^2 - 48m^6$

13. $4x^3 - 4x^2 + x$

14. $5m^4 - 70m^3 + 245m^2$

In Exercises 15–17, show that the binomial is a factor of the polynomial. Then factor the function completely.

15. $f(x) = x^3 - 13x - 12; x + 1$

16. $f(x) = 6x^3 + 8x^2 - 34x - 12; x - 2$

17. $f(x) = 2x^4 - 12x^3 + 6x^2 + 20x; x - 5$

Name ____ Date __________

4.5 Solving Polynomial Equations

For use with Exploration 4.5

Essential Question How can you determine whether a polynomial equation has a repeated solution?

1 EXPLORATION: Cubic Equations and Repeated Solutions

Work with a partner. Some cubic equations have three distinct solutions. Others have repeated solutions. Match each cubic polynomial equation with the graph of its related polynomial function. Then solve each equation. For those equations that have repeated solutions, describe the behavior of the related function near the repeated zero using the graph or a table of values.

a. $x^3 - 6x^2 + 12x - 8 = 0$

b. $x^3 + 3x^2 + 3x + 1 = 0$

c. $x^3 - 3x + 2 = 0$

d. $x^3 + x^2 - 2x = 0$

e. $x^3 - 3x - 2 = 0$

f. $x^3 - 3x^2 + 2x = 0$

A.

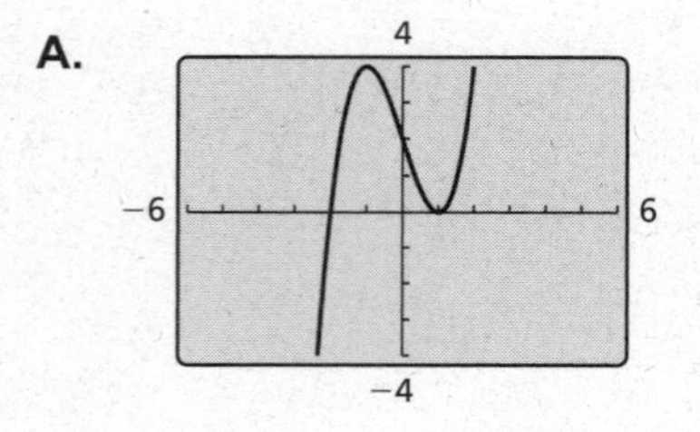

B.

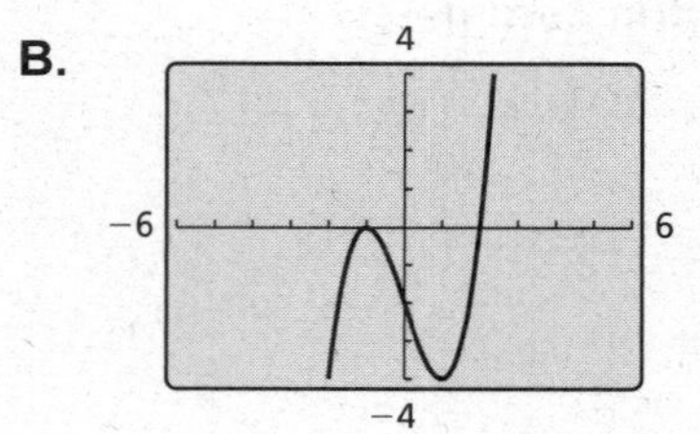

C.

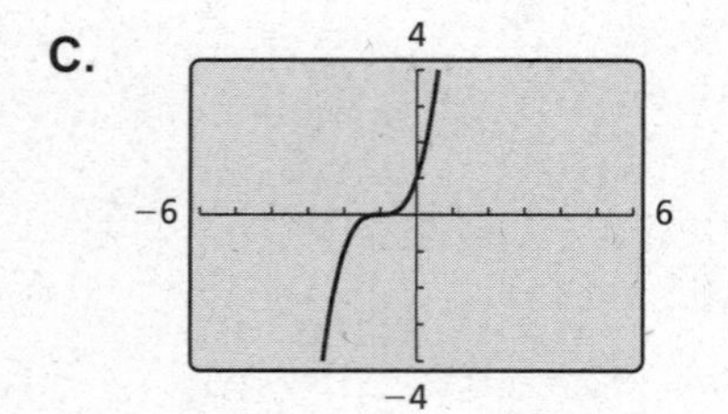

D.

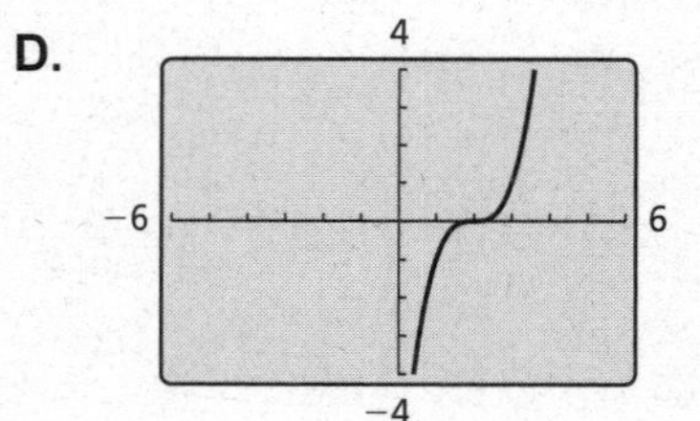

E.

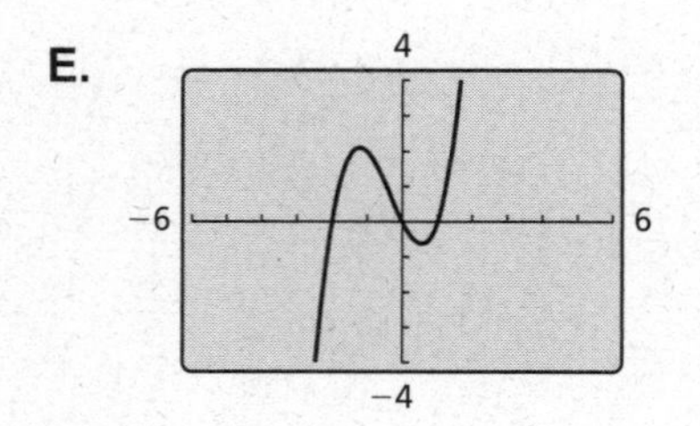

F.

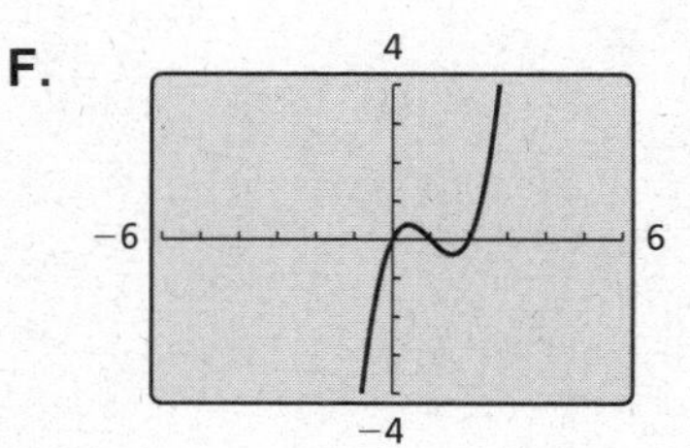

2 EXPLORATION: Quartic Equations and Repeated Solutions

Go to *BigIdeasMath.com* for an interactive tool to investigate this exploration.

Work with a partner. Determine whether each quartic equation has repeated solutions using the graph of the related quartic function or a table of values. Explain your reasoning. Then solve each equation.

a. $x^4 - 4x^3 + 5x^2 - 2x = 0$

b. $x^4 - 2x^3 - x^2 + 2x = 0$

c. $x^4 - 4x^3 + 4x^2 = 0$

d. $x^4 + 3x^3 = 0$

Communicate Your Answer

3. How can you determine whether a polynomial equation has a repeated solution?

4. Write a cubic or a quartic polynomial equation that is different from the equations in Explorations 1 and 2 and has a repeated solution.

Name ______________________________ Date __________

4.5 Notetaking with Vocabulary

For use after Lesson 4.5

In your own words, write the meaning of each vocabulary term.

repeated solution

Core Concepts

The Rational Root Theorem

If $f(x) = a_n x^n + \cdots + a_1 x + a_0$ has *integer* coefficients, then every rational solution of $f(x) = 0$ has the following form:

$$\frac{p}{q} = \frac{\text{factor of constant term } a_0}{\text{factor of leading coefficient } a_n}$$

Notes:

The Irrational Conjugates Theorem

Let f be a polynomial function with rational coefficients, and let a and b be rational numbers such that $\sqrt{b}$ is irrational. If $a + \sqrt{b}$ is a zero of f, then $a - \sqrt{b}$ is also a zero of f.

Notes:

Name______________________________ Date__________

Extra Practice

In Exercises 1–6, solve the equation.

1. $36r^3 - r = 0$

2. $20x^3 + 80x^2 = -60x$

3. $3m^2 = 75m^4$

4. $-13y^2 + 36 = -y^4$

5. $2x^3 - x^2 - 2x = -1$

6. $-20c^2 + 50c = 8c^3 - 125$

In Exercises 7–10, find the zeros of the function. Then sketch a graph of the function.

7. $f(x) = x^4 - x^3 - 12x^2$

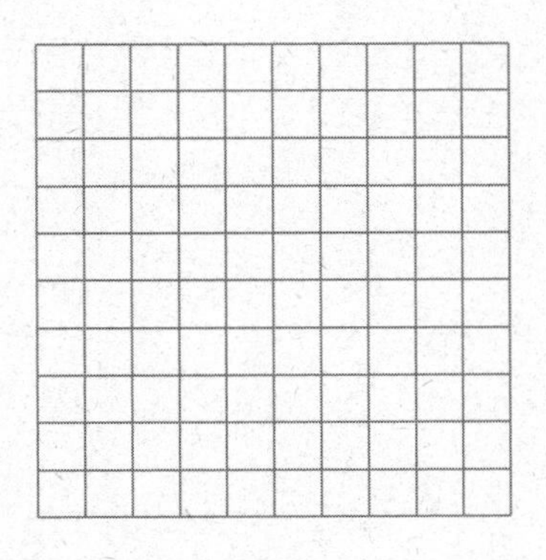

8. $f(x) = -4x^3 + 12x^2 - 9x$

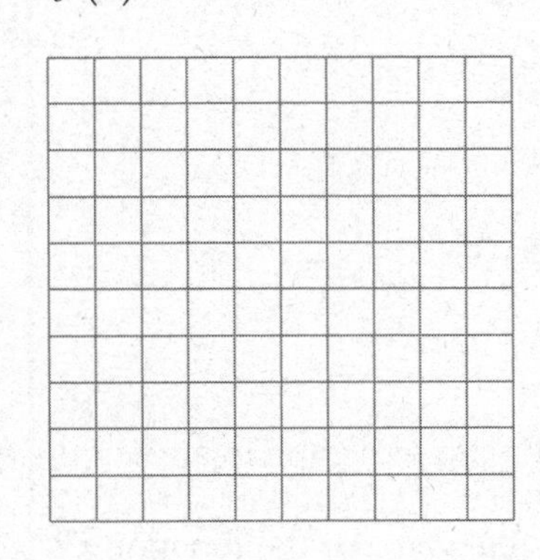

9. $f(x) = x^3 + 4x^2 - 6x - 24$

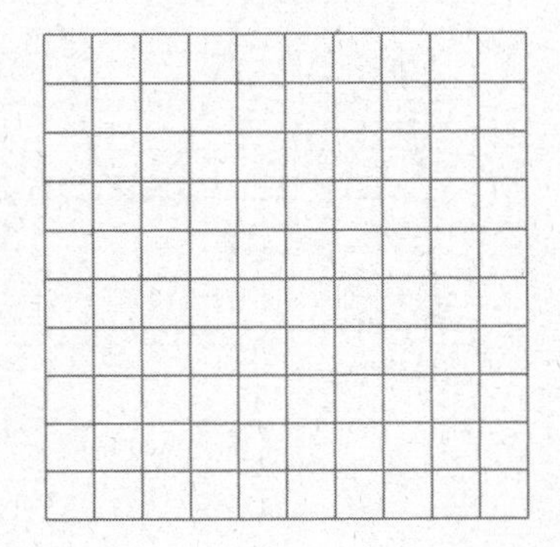

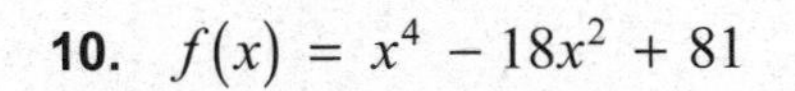

10. $f(x) = x^4 - 18x^2 + 81$

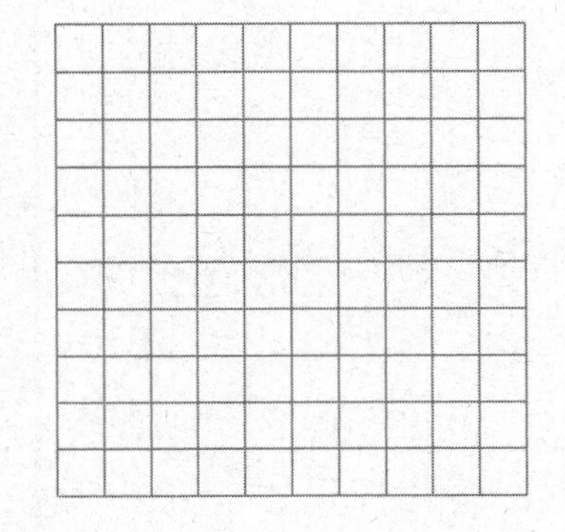

4.5 Notetaking with Vocabulary (continued)

11. According to the Rational Root Theorem, which is *not* a possible solution of the equation $2x^4 + 3x^3 - 6x + 7 = 0$?

A. 3.5 **B.** 0.5 **C.** 7 **D.** 2

12. Find all the real zeros of the function $f(x) = 3x^4 + 11x^3 - 40x^2 - 132x + 48$.

13. Write a polynomial function g of least degree that has rational coefficients, a leading coefficient of 1, and the zeros -5 and $4 + \sqrt{2}$.

14. Use the information in the graph to answer the questions.

a. What are the real zeros of the function f?

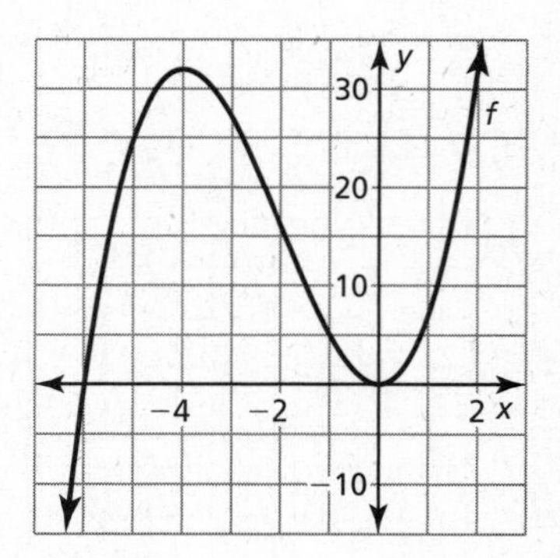

b. Write an equation of the cubic function in factored form.

Name____ Date__________

4.6 The Fundamental Theorem of Algebra

For use with Exploration 4.6

Essential Question How can you determine whether a polynomial equation has imaginary solutions?

1 EXPLORATION: Cubic Equations and Imaginary Solutions

Work with a partner. Match each cubic polynomial equation with the graph of its related polynomial function. Then find *all* solutions. Make a conjecture about how you can use a graph or table of values to determine the number and types of solutions of a cubic polynomial equation.

a. $x^3 - 3x^2 + x + 5 = 0$

b. $x^3 - 2x^2 - x + 2 = 0$

c. $x^3 - x^2 - 4x + 4 = 0$

d. $x^3 + 5x^2 + 8x + 6 = 0$

e. $x^3 - 3x^2 + x - 3 = 0$

f. $x^3 - 3x^2 + 2x = 0$

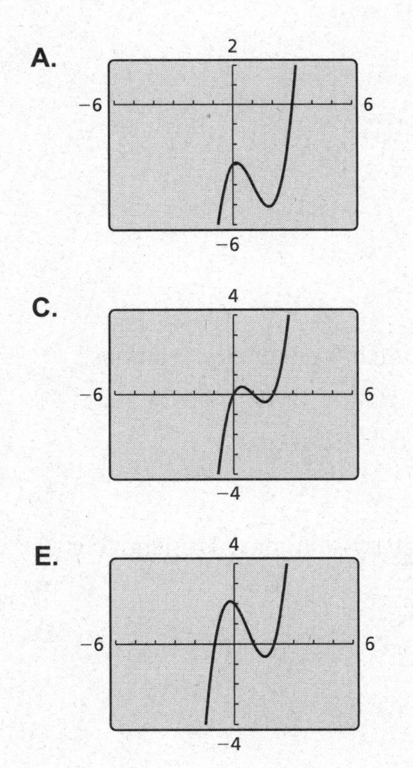

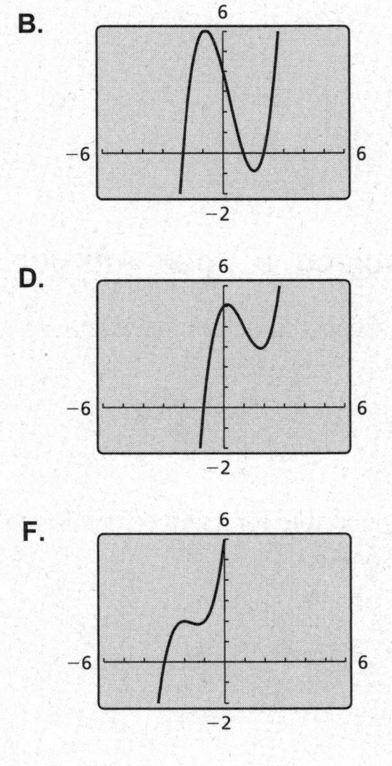

Name ______________________________ Date __________

2 EXPLORATION: Quartic Equations and Imaginary Solutions

Go to *BigIdeasMath.com* for an interactive tool to investigate this exploration.

Work with a partner. Use the graph of the related quartic function, or a table of values, to determine whether each quartic equation has imaginary solutions. Explain your reasoning. Then find *all* solutions.

a. $x^4 - 2x^3 - x^2 + 2x = 0$

b. $x^4 - 1 = 0$

c. $x^4 + x^3 - x - 1 = 0$

d. $x^4 - 3x^3 + x^2 + 3x - 2 = 0$

Communicate Your Answer

3. How can you determine whether a polynomial equation has imaginary solutions?

4. Is it possible for a cubic equation to have three imaginary solutions? Explain your reasoning.

Name______________________________ Date__________

4.6 Notetaking with Vocabulary

For use after Lesson 4.6

In your own words, write the meaning of each vocabulary term.

complex conjugates

Core Concepts

The Fundamental Theorem of Algebra

Theorem If $f(x)$ is a polynomial of degree n where $n > 0$, then the equation $f(x) = 0$ has at least one solution in the set of complex numbers.

Corollary If $f(x)$ is a polynomial of degree n where $n > 0$, then the equation $f(x) = 0$ has exactly n solutions provided each solution repeated twice is counted as two solutions, each solution repeated three times is counted as three solutions, and so on.

Notes:

The Complex Conjugates Theorem

If *f* is a polynomial function with real coefficients, and $a + bi$ is an imaginary zero of *f*, then $a - bi$ is also a zero of *f*.

Notes:

Descartes's Rule of Signs

Let $f(x) = a_n x^n + a_{n-1}x^{n-1} + \cdots + a_2 x^2 + a_1 x + a_0$ be a polynomial function with real coefficients.

- The number of *positive real zeros* of *f* is equal to the number of changes in sign of the coefficients of $f(x)$ or is less than this by an even number.
- The number of *negative real zeros* of *f* is equal to the number of changes in sign of the coefficients of $f(-x)$ or is less than this by an even number.

Notes:

Name ____ Date __________

Extra Practice

In Exercises 1–4, find all zeros of the polynomial function.

1. $h(x) = x^4 - 3x^3 + 6x^2 + 2x - 60$

2. $f(x) = x^3 - 3x^2 - 15x + 125$

3. $g(x) = x^4 - 48x^2 - 49$

4. $h(x) = -5x^3 + 9x^2 - 18x - 4$

In Exercises 5–8, write a polynomial function *f* of least degree that has rational coefficients, a leading coefficient of 1, and the given zeros.

5. $-4, 1, 7$

6. $10, -\sqrt{5}$

7. $8, 3 - i$

8. $0, 2 - \sqrt{2}, 2 + 3i$

Name ______________________________ Date ________

4.7 Transformations of Polynomial Functions

For use with Exploration 4.7

Essential Question How can you transform the graph of a polynomial function?

1 EXPLORATION: Transforming the Graph of the Cubic Function

Go to *BigIdeasMath.com* for an interactive tool to investigate this exploration.

Work with a partner. The graph of the cubic function

$$f(x) = x^3$$

is shown. The graph of each cubic function g represents a transformation of the graph of f. Write a rule for g. Use a graphing calculator to verify your answers.

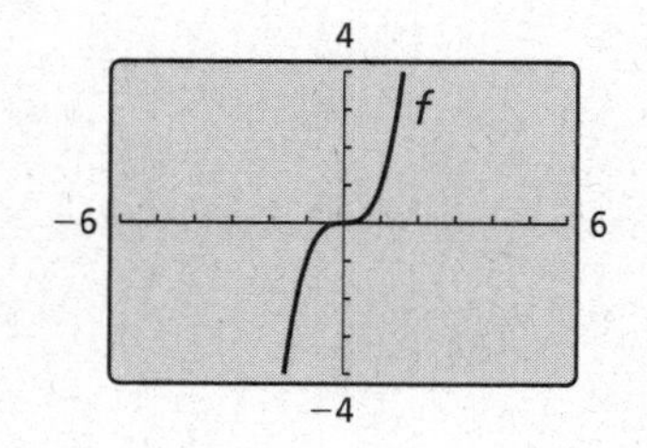

a.

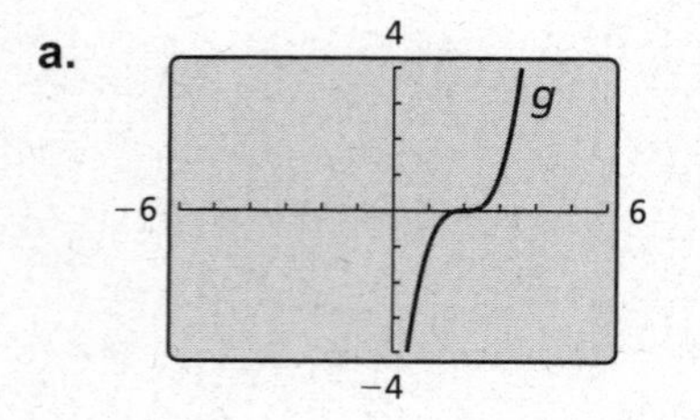

b.

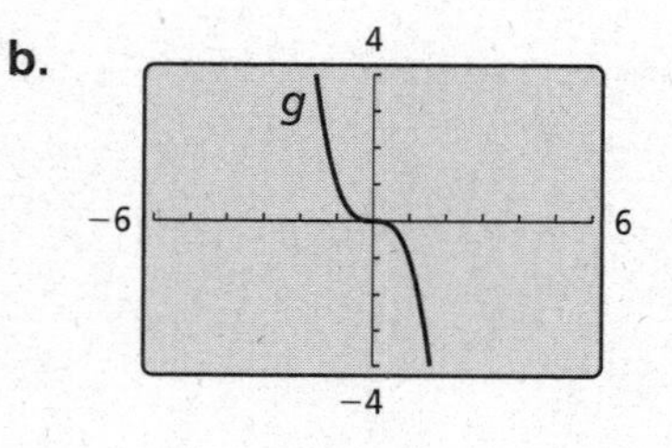

c.

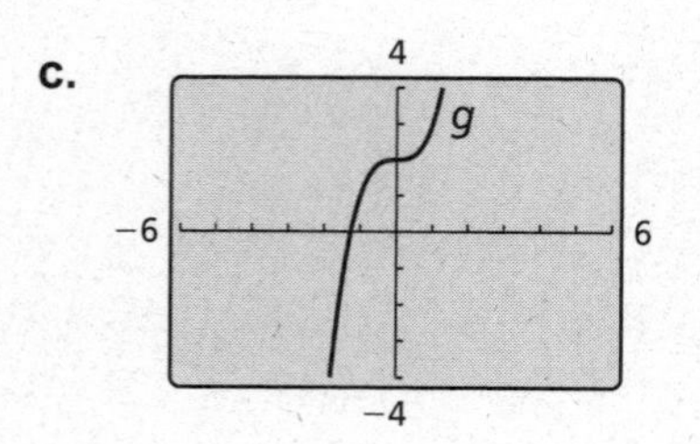

d.

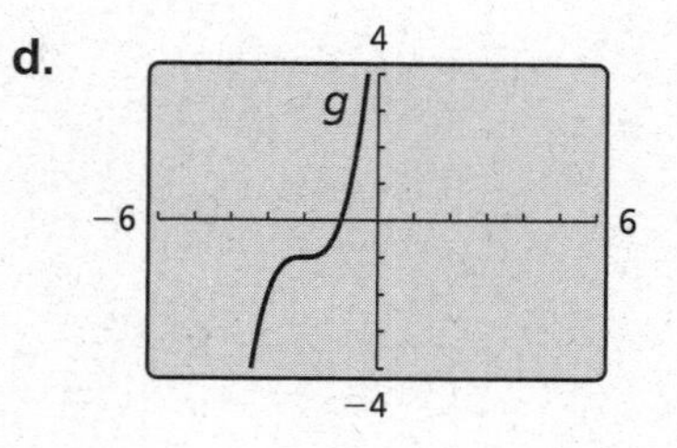

Name____ Date__________

2 EXPLORATION: Transforming the Graph of the Quartic Function

Go to *BigIdeasMath.com* for an interactive tool to investigate this exploration.

Work with a partner. The graph of the quartic function

$f(x) = x^4$

is shown. The graph of each quartic function g represents a transformation of the graph of f. Write a rule for g. Use a graphing calculator to verify your answers.

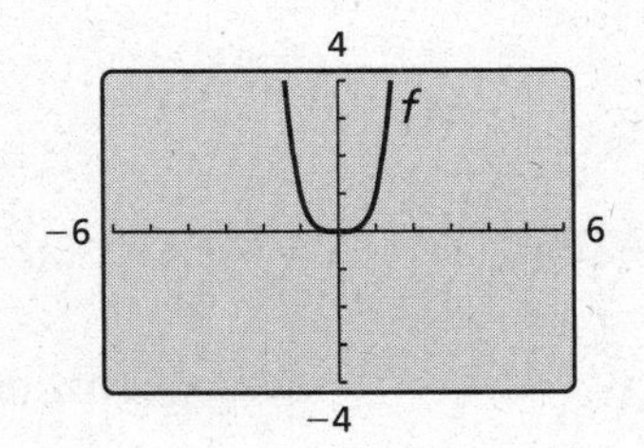

a.

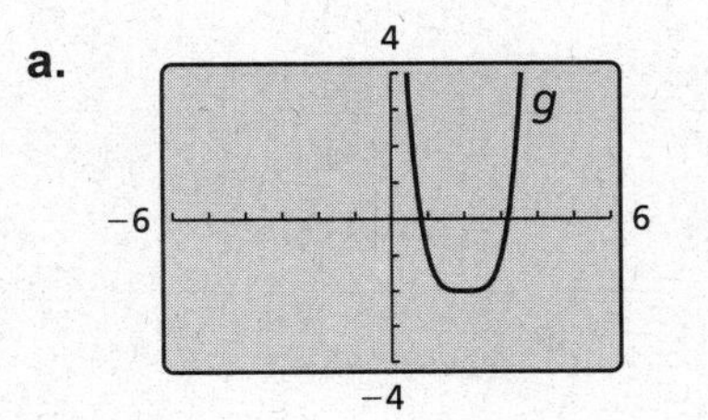

b.

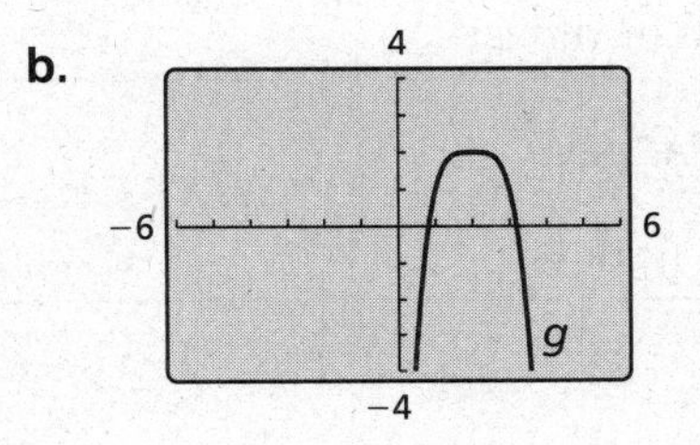

Communicate Your Answer

3. How can you transform the graph of a polynomial function?

4. Describe the transformation of $f(x) = x^4$ represented by $g(x) = (x + 1)^4 + 3$. Then graph $g(x)$.

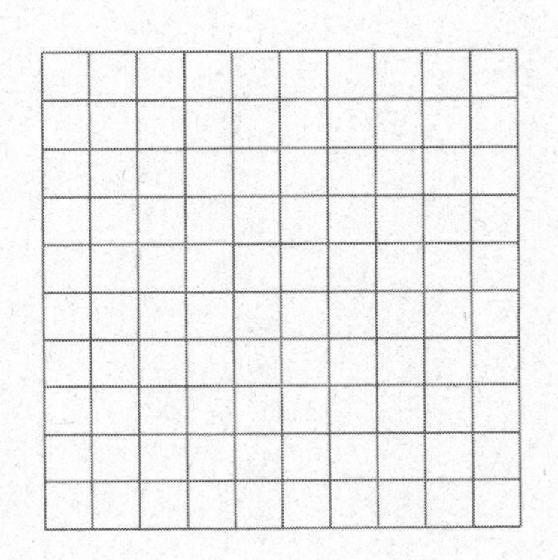

Name ____ Date __________

4.7 Notetaking with Vocabulary

For use after Lesson 4.7

In your own words, write the meaning of each vocabulary term.

polynomial function

transformations

Core Concepts

Transformation	$f(x)$ Notation	Examples	
Horizontal Translation Graph shifts left or right.	$f(x - h)$	$g(x) = (x - 5)^4$ $g(x) = (x + 2)^4$	5 units right 2 units left
Vertical Translation Graph shifts up or down.	$f(x) + k$	$g(x) = x^4 + 1$ $g(x) = x^4 - 4$	1 unit up 4 units down
Reflection Graph flips over x- or y-axis.	$f(-x)$ $-f(x)$	$g(x) = (-x)^4 = x^4$ $g(x) = -x^4$	over y-axis over x-axis
Horizontal Stretch or Shrink Graph stretches away from or shrinks toward y-axis	$f(ax)$	$g(x) = (2x)^4$ $g(x) = \left(\frac{1}{2}x\right)^4$	shrink by $\frac{1}{2}$ stretch by 2
Vertical Stretch or Shrink Graph stretches away from or shrinks toward x-axis.	$a \bullet f(x)$	$g(x) = 8x^4$ $g(x) = \frac{1}{4}x^4$	stretch by 8 shrink by $\frac{1}{4}$

Notes:

Extra Practice

In Exercises 1–6, describe the transformation of *f* represented by *g*. Then graph each function.

1. $f(x) = x^4;\ g(x) = x^4 - 9$

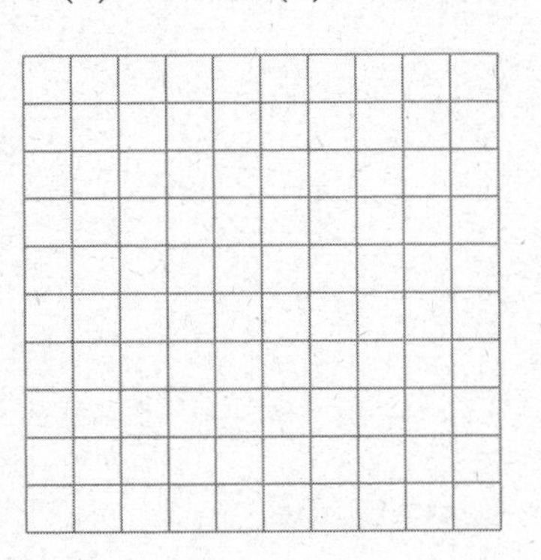

2. $f(x) = x^5;\ g(x) = (x + 1)^5 + 2$

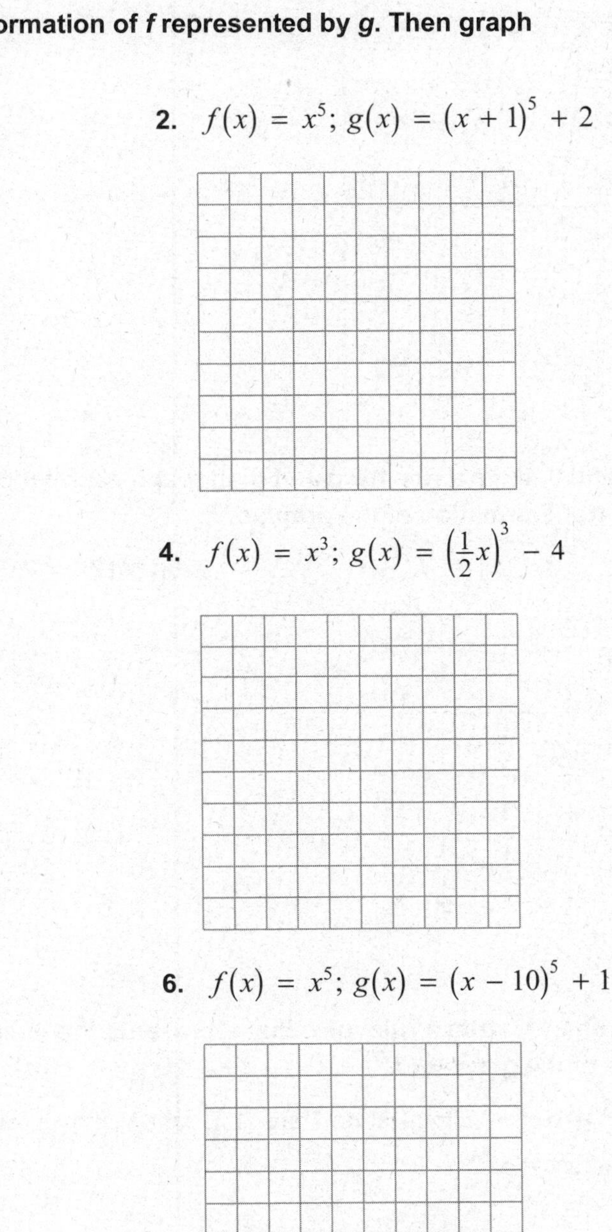

3. $f(x) = x^6;\ g(x) = -5(x - 2)^6$

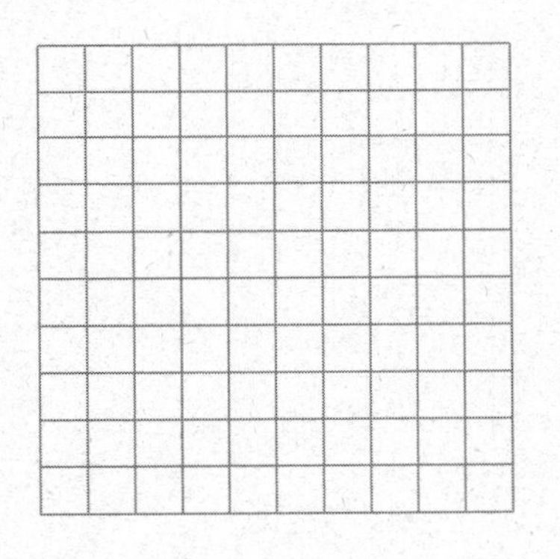

4. $f(x) = x^3;\ g(x) = \left(\frac{1}{2}x\right)^3 - 4$

5. $f(x) = x^4;\ g(x) = \frac{1}{8}(-x)^4$

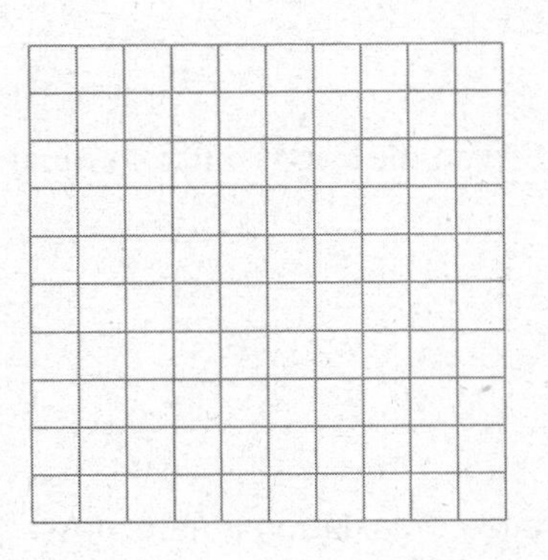

6. $f(x) = x^5;\ g(x) = (x - 10)^5 + 1$

Name ______________________________ Date __________

4.7 Notetaking with Vocabulary (continued)

7. Graph the function $g(x) = -f(x - 3)$ on the same coordinate plane as $f(x)$.

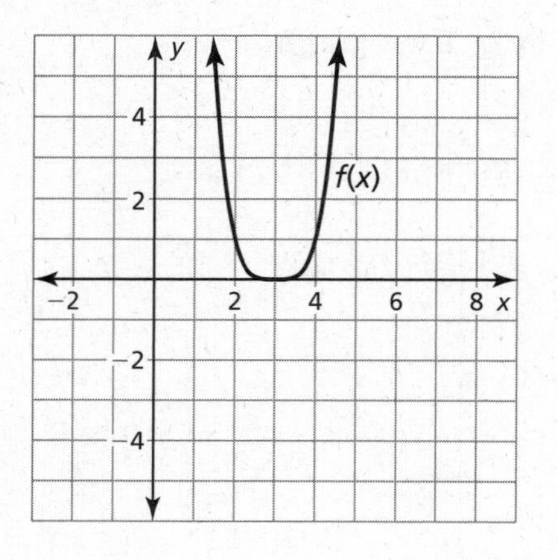

In Exercises 8 and 9, write a rule for *g* and then graph each function. Describe the graph of *g* as a transformation of the graph of *f*.

8. $f(x) = x^3 + 8; g(x) = f(-x) - 9$

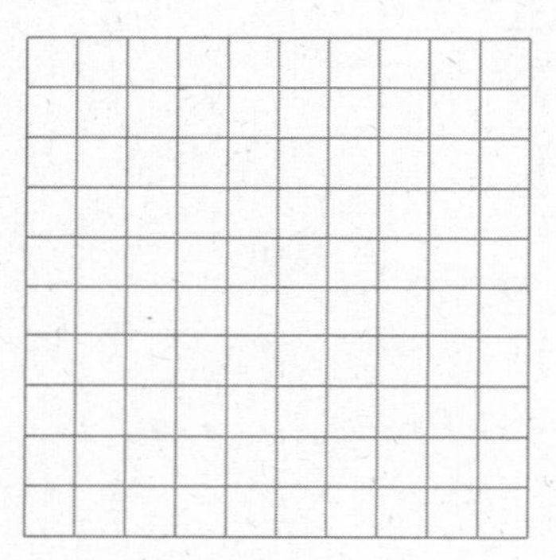

9. $f(x) = 2x^5 - x^3 + 1; g(x) = 5f(x)$

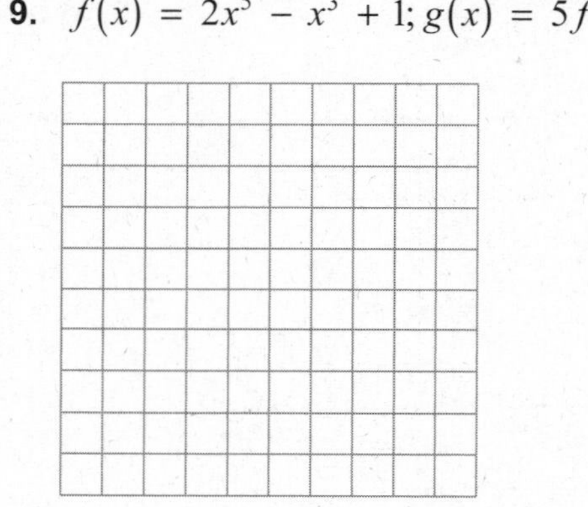

In Exercises 10 and 11, write a rule for *g* that represents the indicated transformations of the graph of *f*.

10. $f(x) = x^3 - 6x^2 + 5$; translation 1 unit left, followed by a reflection in the x-axis and a vertical stretch by a factor of 2

11. $f(x) = 3x^4 + x^3 + 3x^2 + 12$; horizontal shrink by a factor of $\frac{1}{3}$ and a translation 8 units down, followed by a reflection in the y-axis

Name____ Date__________

4.8 Analyzing Graphs of Polynomial Functions

For use with Exploration 4.8

Essential Question How many turning points can the graph of a polynomial function have?

1 EXPLORATION: Approximating Turning Points

Go to *BigIdeasMath.com* for an interactive tool to investigate this exploration.

Work with a partner. Match each polynomial function with its graph. Explain your reasoning. Then use a graphing calculator to approximate the coordinates of the turning points of the graph of the function. Round your answers to the nearest hundredth.

a. $f(x) = 2x^2 + 3x - 4$

b. $f(x) = x^2 + 3x + 2$

c. $f(x) = x^3 - 2x^2 - x + 1$

d. $f(x) = -x^3 + 5x - 2$

e. $f(x) = x^4 - 3x^2 + 2x - 1$

f. $f(x) = -2x^5 - x^2 + 5x + 3$

A.

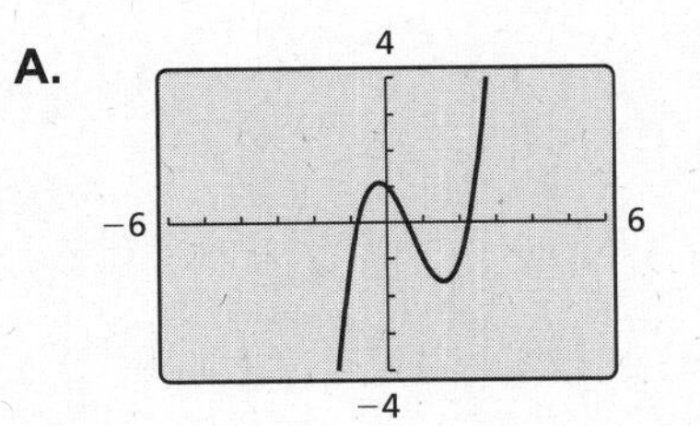

B.

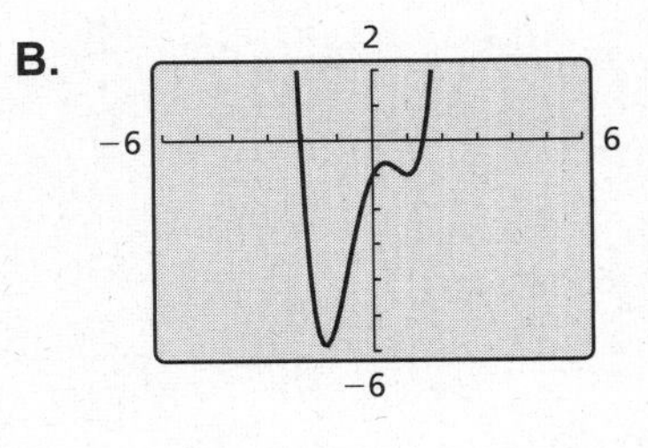

C.

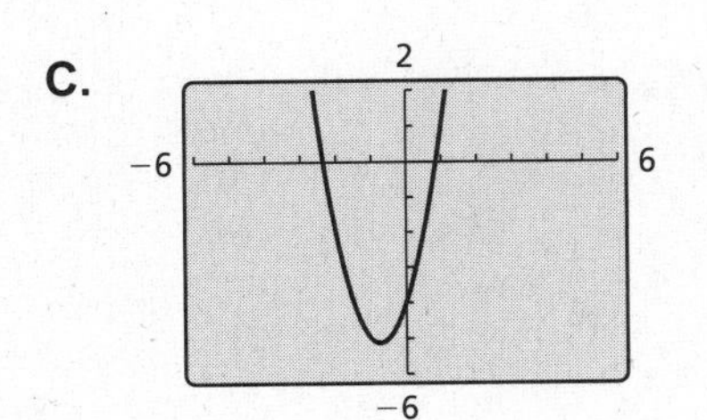

D.

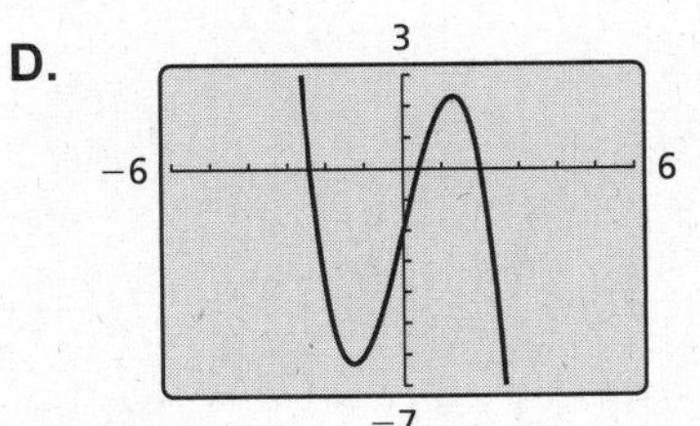

E.

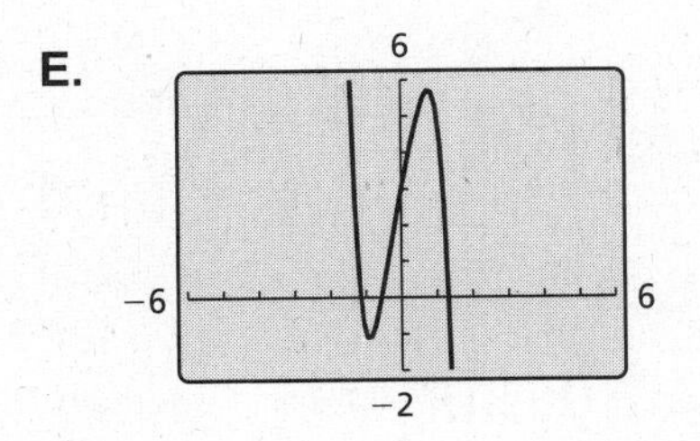

F.

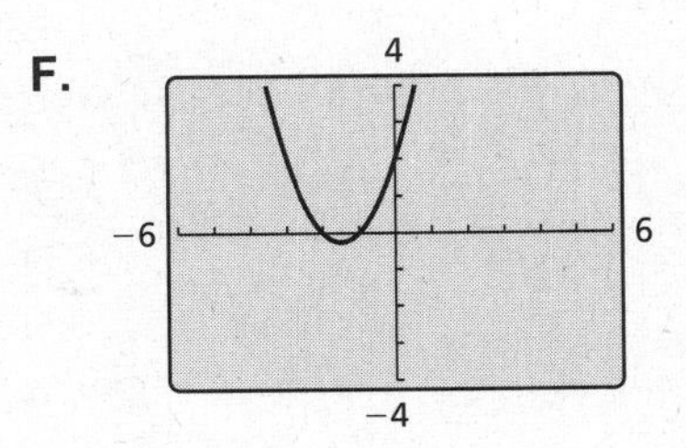

Communicate Your Answer

2. How many turning points can the graph of a polynomial function have?

3. Is it possible to sketch the graph of a cubic polynomial function that has *no* turning points? Justify your answer.

Name______________________________ Date__________

4.8 Notetaking with Vocabulary

For use after Lesson 4.8

In your own words, write the meaning of each vocabulary term.

local maximum

local minimum

even function

odd function

Core Concepts

Zeros, Factors, Solutions, and Intercepts

Let $f(x) = a_n x^n + a_{n-1} x^{n-1} + \cdots + a_1 x + a_0$ be a polynomial function. The following statements are equivalent.

Zero: k is a zero of the polynomial function f.

Factor: $x - k$ is a factor of the polynomial $f(x)$.

Solution: k is a solution (or root) of the polynomial equation $f(x) = 0$.

x-Intercept: If k is a real number, then k is an x-intercept of the graph of the polynomial function f. The graph of f passes through $(k, 0)$.

Notes:

The Location Principle

If f is a polynomial function, and a and b are two real numbers such that $f(a) < 0$ and $f(b) > 0$, then f has at least one real zero between a and b.

Notes:

Turning Points of Polynomial Functions

1. The graph of every polynomial function of degree n has *at most* $n - 1$ turning points.
2. If a polynomial function has n distinct real zeros, then its graph has *exactly* $n - 1$ turning points.

Notes:

Even and Odd Functions

A function f is an **even function** when $f(-x) = f(x)$ for all x in its domain. The graph of an even function is *symmetric about the y-axis.*

A function f is an **odd function** when $f(-x) = -f(x)$ for all x in its domain. The graph of an odd function is *symmetric about the origin.* One way to recognize a graph that is symmetric about the origin is that it looks the same after a $180°$ rotation about the origin.

Even Function

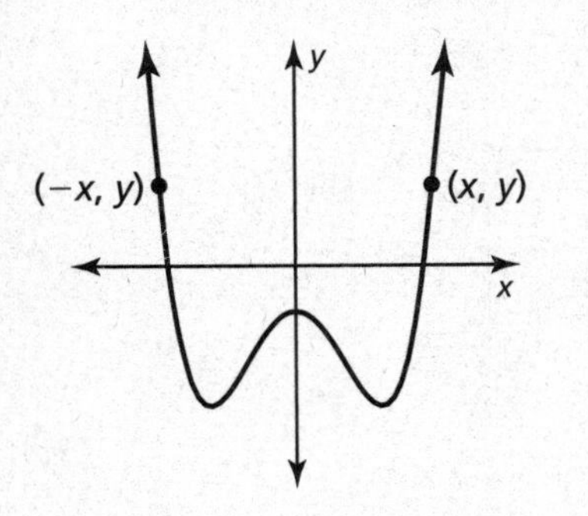

For an even function, if (x, y) is on the graph, then $(-x, y)$ is also on the graph.

Odd Function

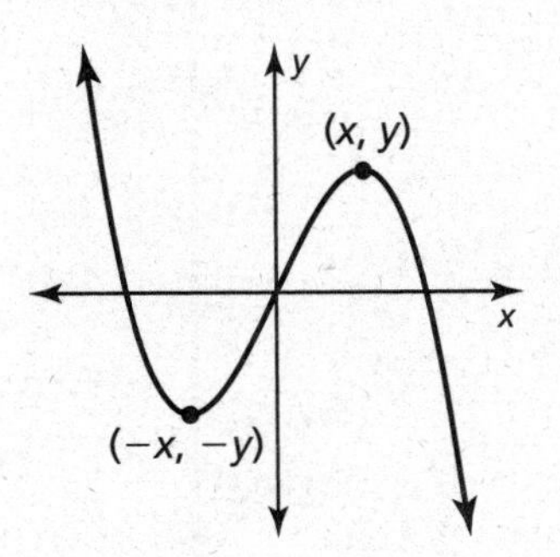

For an odd function, if (x, y) is on the graph, then $(-x, -y)$ is also on the graph.

Notes:

Name ______________________________ Date __________

Extra Practice

In Exercises 1–6, graph the function. Identify the *x*-intercepts, and the points where the local maximums and local minimums occur. Determine the intervals for which the function is increasing or decreasing. Determine whether the function is *even*, *odd*, or *neither*.

1. $f(x) = 4x^3 - 12x^2 - x + 15$

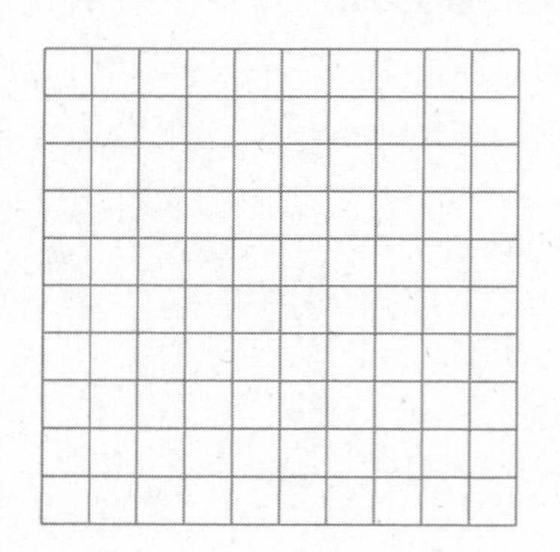

2. $g(x) = 2x^4 + 5x^3 - 21x^2 - 10x$

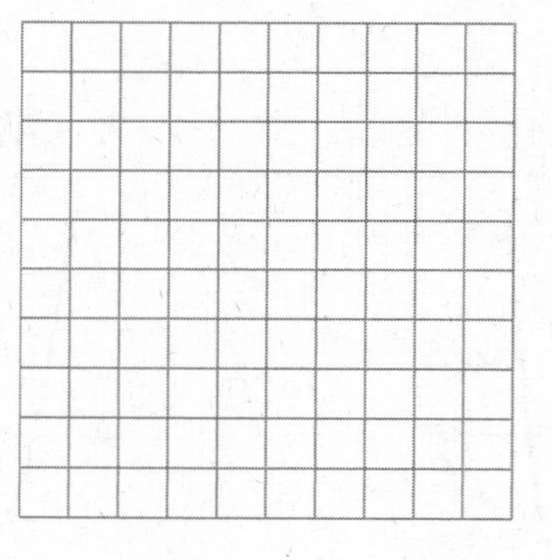

3. $h(x) = x^3 - x^2 - 13x - 3$

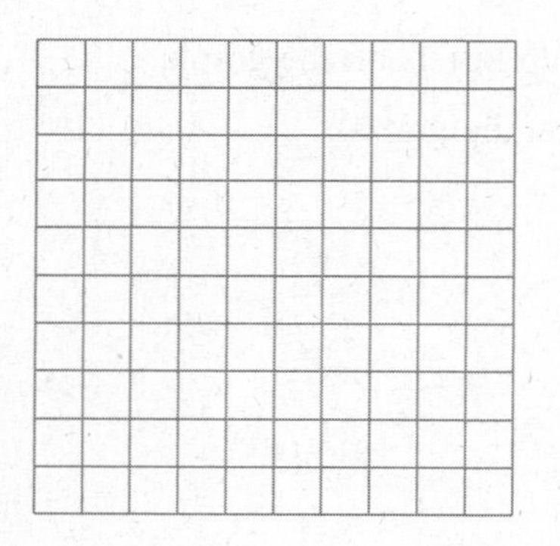

4. $k(x) = x^3 - 2x$

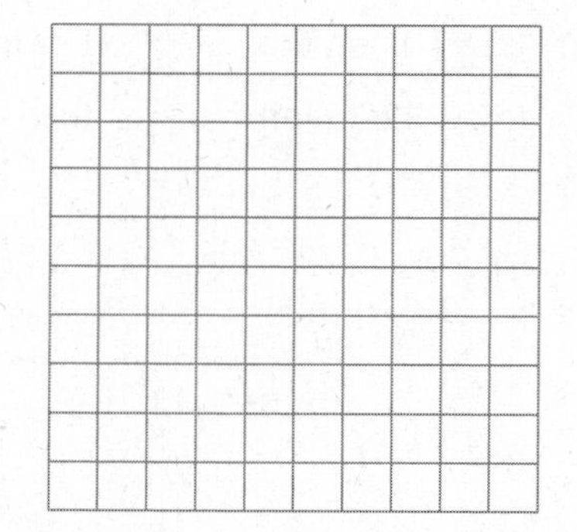

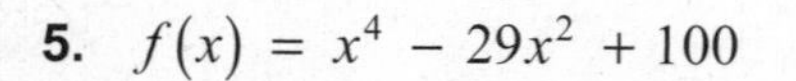

5. $f(x) = x^4 - 29x^2 + 100$

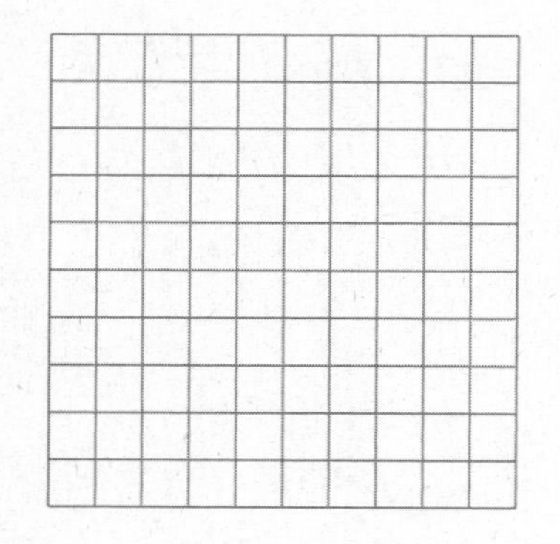

6. $g(x) = -\frac{1}{3}x^3 + x^2 - \frac{4}{3}$

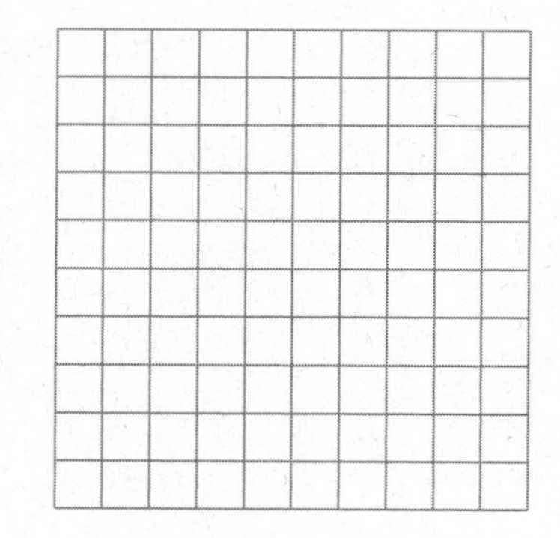

Name______________________________ Date__________

4.9 Modeling with Polynomial Functions

For use with Exploration 4.9

Essential Question How can you find a polynomial model for real-life data?

1 EXPLORATION: Modeling Real-Life Data

Go to *BigIdeasMath.com* for an interactive tool to investigate this exploration.

Work with a partner. The distance a baseball travels after it is hit depends on the angle at which it was hit and the initial speed. The table shows the distances a baseball hit at an angle of 35° travels at various initial speeds.

Initial speed, *x* (miles per hour)	80	85	90	95	100	105	110	115
Distance, *y* (feet)	194	220	247	275	304	334	365	397

a. Recall that when data have equally-spaced x-values, you can analyze patterns in the differences of the y-values to determine what type of function can be used to model the data. If the first differences are constant, then the set of data fits a linear model. If the second differences are constant, then the set of data fits a quadratic model.

Find the first and second differences of the data. Are the data linear or quadratic? Explain your reasoning.

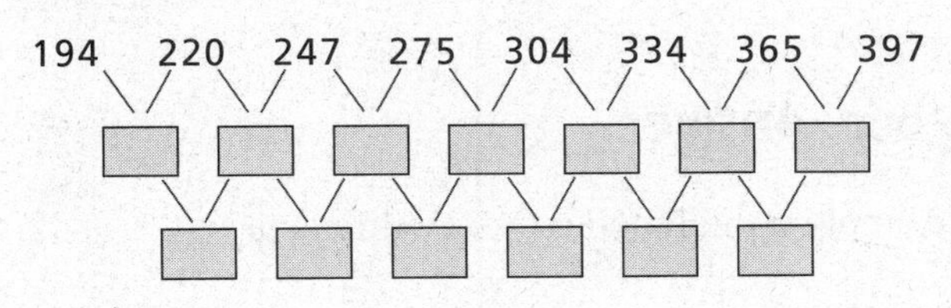

b. Use a graphing calculator to draw a scatter plot of the data. Do the data appear linear or quadratic? Use the *regression* feature of the graphing calculator to find a linear or quadratic model that best fits the data.

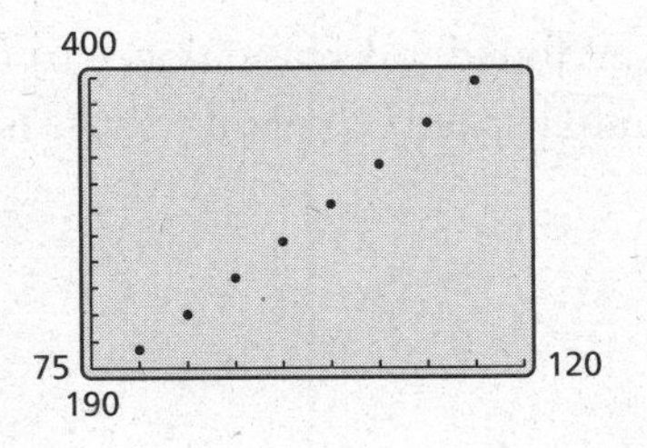

4.9 Modeling with Polynomial Functions (continued)

1 EXPLORATION: Modeling Real-Life Data (continued)

c. Use the model you found in part (b) to find the distance a baseball travels when it is hit at an angle of 35° and travels at an initial speed of 120 miles per hour.

d. According to the *Baseball Almanac,* "Any drive over 400 feet is noteworthy. A blow of 450 feet shows exceptional power, as the majority of major league players are unable to hit a ball that far. Anything in the 500-foot range is genuinely historic." Estimate the initial speed of a baseball that travels a distance of 500 feet.

Communicate Your Answer

2. How can you find a polynomial model for real-life data?

3. How well does the model you found in Exploration 1(b) fit the data? Do you think the model is valid for any initial speed? Explain your reasoning.

Name ______________________________ Date __________

4.9 Notetaking with Vocabulary

For use after Lesson 4.9

In your own words, write the meaning of each vocabulary term.

finite differences

Core Concepts

Properties of Finite Differences

1. If a polynomial function $f(x)$ has degree n, then the nth differences of function values for equally-spaced x-values are nonzero and constant.

2. Conversely, if the nth differences of equally-spaced data are nonzero and constant, then the data can be represented by a polynomial function of degree n.

Notes:

Name ______________________________ Date __________

Extra Practice

In Exercises 1–4, write a cubic function whose graph passes through the given points.

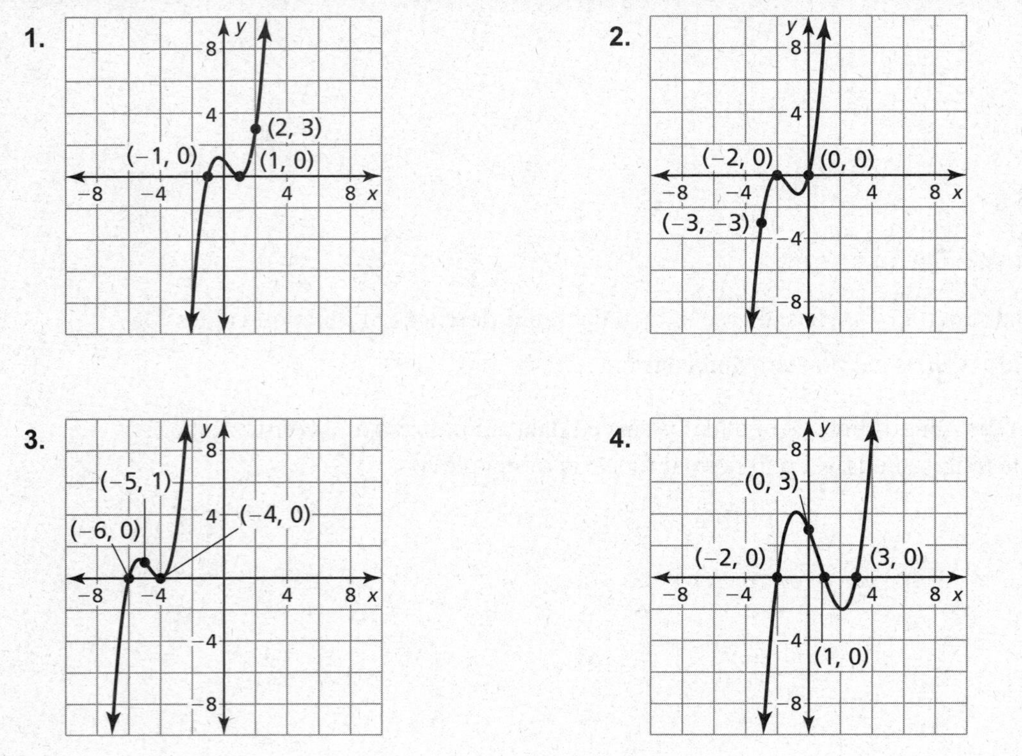

In Exercises 5–8, use finite differences to determine the degree of the polynomial function that fits the data. Then use technology to find the polynomial function.

5.

x	−2	−1	0	1	2	3
$f(x)$	−14	−6.5	0	5.5	10	13.5

6.

x	−2	−1	0	1	2
$f(x)$	30	4	0	0	−14

7. $(0, 0), (2, 0), (4, 40), (6, 168), (8, 432), (10, 880)$

8. $(0, 10), (1, 10), (2, 18), (3, 64), (4, 202), (5, 510)$

4.9 Notetaking with Vocabulary (continued)

9. The table shows the population y of bacteria after x hours. Find a polynomial model for the data for the first 4.5 hours. Use the model to estimate the population level of the bacteria (in thousands) after 1 day.

Number of Hours, *x*	0.5	1	2.5	3	4	4.5
Number of Bacteria, *y*	5.125	6	20.625	32	69	96.125

10. The table shows the value y (in thousands of dollars) of a signed autograph of a MVP football player, where x represents the number of years since 2000. Find a polynomial model for the data for the first 5 years. Use the model to estimate the year that the autograph will be valued at $5,000,000?

Number of Years, *x*	1	2	3	4	5
Value of autograph, *y*	6	34	162	510	1246

Name ______________________ Date __________

Chapter 5 Maintaining Mathematical Proficiency

Simplify the expression.

1. $c \bullet c^9$

2. $\dfrac{q^{12}}{q^4}$

3. $\dfrac{x^3}{x^4 \bullet x^5}$

4. $\dfrac{d^2}{d} \bullet 8d^5$

5. $\left(\dfrac{4x^3}{2y^4}\right)^2$

6. $\left(\dfrac{m^8 \bullet m^3}{n \bullet m}\right)^3$

Solve the literal equation for *y*.

7. $x + y = 1$

8. $-3y + \frac{1}{2}x = -6$

9. $24x + 5y = 74$

10. $6xy + 3y = -72$

11. $10x - 5xy = 100$

12. $-\frac{1}{4}x + 8xy = 16$

13. Is $\left(\dfrac{x + 3x}{y + 2y}\right)^2 = \left(\dfrac{x^2 + 3^2x^2}{y^2 + 2^2y^2}\right)$ or is $\left(\dfrac{x + 3x}{y + 2y}\right)^2 = \left(\dfrac{4^2x^2}{3^2y^2}\right)$? Explain your reasoning.

Name____ Date__________

5.1 *n*th Roots and Rational Exponents

For use with Exploration 5.1

Essential Question How can you use a rational exponent to represent a power involving a radical?

1 EXPLORATION: Exploring the Definition of a Rational Exponent

Go to *BigIdeasMath.com* for an interactive tool to investigate this exploration.

Work with a partner. Use a calculator to show that each statement is true.

a. $\sqrt{9} = 9^{1/2}$

b. $\sqrt{2} = 2^{1/2}$

c. $\sqrt[3]{8} = 8^{1/3}$

d. $\sqrt[3]{3} = 3^{1/3}$

e. $\sqrt[4]{16} = 16^{1/4}$

f. $\sqrt[4]{12} = 12^{1/4}$

2 EXPLORATION: Writing Expressions in Rational Exponent Form

Go to *BigIdeasMath.com* for an interactive tool to investigate this exploration.

Work with a partner. Use the definition of a rational exponent and the properties of exponents to write each expression as a base with a single rational exponent. Then use a calculator to evaluate each expression. Round your answer to two decimal places.

Sample

$$\left(\sqrt[3]{4}\right)^2 = \left(4^{1/3}\right)^2$$
$$= 4^{2/3}$$
$$\approx 2.52$$

```
4^(2/3)
               2.5198421
```

a. $\left(\sqrt{5}\right)^3$

b. $\left(\sqrt[4]{4}\right)^2$

c. $\left(\sqrt[3]{9}\right)^2$

d. $\left(\sqrt[5]{10}\right)^4$

e. $\left(\sqrt{15}\right)^3$

f. $\left(\sqrt[3]{27}\right)^4$

3 EXPLORATION: Writing Expressions in Radical Form

Go to *BigIdeasMath.com* for an interactive tool to investigate this exploration.

Work with a partner. Use the properties of exponents and the definition of a rational exponent to write each expression as a radical raised to an exponent. Then use a calculator to evaluate each expression. Round your answer to two decimal places.

Sample $5^{2/3} = \left(5^{1/3}\right)^2 = \left(\sqrt[3]{5}\right)^2 \approx 2.92$

a. $8^{2/3}$

b. $6^{5/2}$

c. $12^{3/4}$

d. $10^{3/2}$

e. $16^{3/2}$

f. $20^{6/5}$

Communicate Your Answer

4. How can you use a rational exponent to represent a power involving a radical?

5. Evaluate each expression *without* using a calculator. Explain your reasoning.

a. $4^{3/2}$

b. $32^{4/5}$

c. $625^{3/4}$

d. $49^{3/2}$

e. $125^{4/3}$

f. $100^{6/3}$

Name____ Date__________

5.1 Notetaking with Vocabulary

For use after Lesson 5.1

In your own words, write the meaning of each vocabulary term.

*n*th root of *a*

index of a radical

Core Concepts

Real *n*th roots of *a*

Let n be an integer $(n > 1)$ and let a be a real number.

***n* is an even integer.**	***n* is an odd integer.**
$\boldsymbol{a < 0}$ No real *n*th roots	$\boldsymbol{a < 0}$ One real *n*th root: $\sqrt[n]{a} = a^{1/n}$
$\boldsymbol{a = 0}$ One real *n*th root: $\sqrt[n]{0} = 0$	$\boldsymbol{a = 0}$ One real *n*th root: $\sqrt[n]{0} = 0$
$\boldsymbol{a > 0}$ Two real *n*th roots: $\pm\sqrt[n]{a} = \pm a^{1/n}$	$\boldsymbol{a > 0}$ One real *n*th root: $\sqrt[n]{a} = a^{1/n}$

Notes:

Name ______________________________ Date __________

Rational Exponents

Let $a^{1/n}$ be an nth root of a, and let m be a positive integer.

$$a^{m/n} = \left(a^{1/n}\right)^m = \left(\sqrt[n]{a}\right)^m$$

$$a^{-m/n} = \frac{1}{a^{m/n}} = \frac{1}{\left(a^{1/n}\right)^m} = \frac{1}{\left(\sqrt[n]{a}\right)^m}, a \neq 0$$

Notes:

Extra Practice

In Exercises 1–3, find the indicated real *n*th root(s) of *a*.

1. $n = 3, a = -125$ **2.** $n = 2, a = -400$ **3.** $n = 6, a = 64$

In Exercises 4–11, evaluate the expression without using a calculator.

4. $64^{1/2}$ **5.** $(-27)^{1/3}$ **6.** $32^{7/5}$ **7.** $49^{-3/2}$

8. $(-32)^{3/5}$ **9.** $1000^{-2/3}$ **10.** $81^{3/4}$ **11.** $625^{1/4}$

5.1 Notetaking with Vocabulary (continued)

In Exercises 12–15, match the equivalent expressions. Explain your reasoning.

12. $\left(\sqrt{a}\right)^3$

13. $-\sqrt[3]{a}$

14. $\left(\sqrt[3]{a}\right)^2$

15. $\dfrac{1}{\sqrt[3]{a}}$

A. $a^{-1/3}$

B. $a^{2/3}$

C. $a^{3/2}$

D. $-a^{1/3}$

In Exercises 16–19, find the real solution(s) of the equation. Round your answer to two decimal places when appropriate.

16. $6x^3 = -6$

17. $2(x + 5)^4 = 128$

18. $x^5 - 32 = -64$

19. $-\frac{1}{10}x^3 + 100 = 0$

20. The volume of a cube is 1728 cubic inches. What are the dimensions of the cube?

Name ____________________ Date ________

5.2 Properties of Rational Exponents and Radicals
For use with Exploration 5.2

Essential Question How can you use properties of exponents to simplify products and quotients of radicals?

1 EXPLORATION: Reviewing Properties of Exponents

Work with a partner. Let *a* and *b* be real numbers. Use the properties of exponents to complete each statement. Then match each completed statement with the property it illustrates.

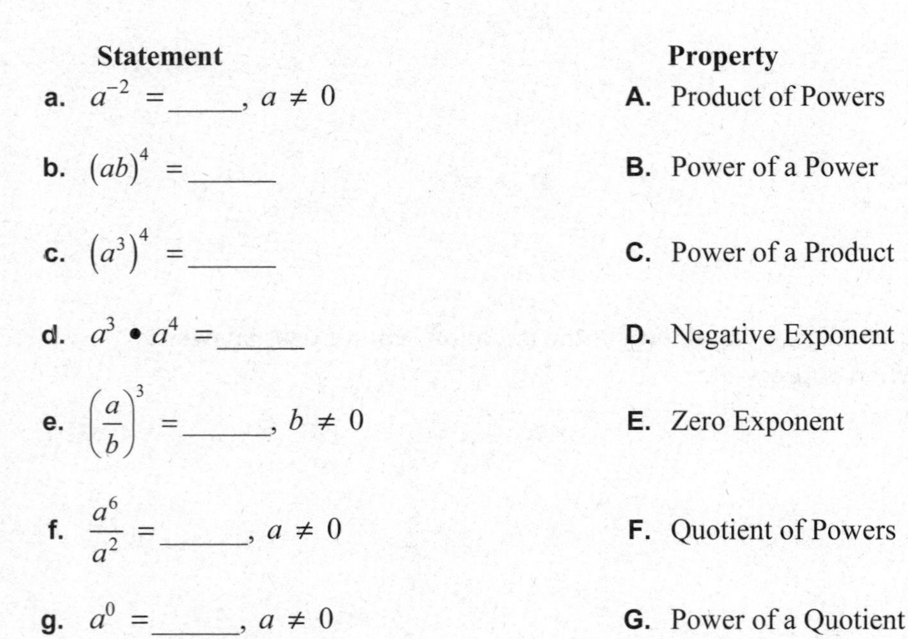

Statement	Property
a. $a^{-2} = \_\_\_\_\_, a \neq 0$	**A.** Product of Powers
b. $(ab)^4 = \_\_\_\_\_$	**B.** Power of a Power
c. $(a^3)^4 = \_\_\_\_\_$	**C.** Power of a Product
d. $a^3 \bullet a^4 = \_\_\_\_\_$	**D.** Negative Exponent
e. $\left(\frac{a}{b}\right)^3 = \_\_\_\_\_, b \neq 0$	**E.** Zero Exponent
f. $\frac{a^6}{a^2} = \_\_\_\_\_, a \neq 0$	**F.** Quotient of Powers
g. $a^0 = \_\_\_\_\_, a \neq 0$	**G.** Power of a Quotient

2 EXPLORATION: Simplifying Expressions with Rational Exponents

Go to *BigIdeasMath.com* for an interactive tool to investigate this exploration.

Work with a partner. Show that you can apply the properties of integer exponents to rational exponents by simplifying each expression. Use a calculator to check your answers.

a. $5^{2/3} \bullet 5^{4/3}$ **b.** $3^{1/5} \bullet 3^{4/5}$ **c.** $(4^{2/3})^3$

d. $(10^{1/2})^4$ **e.** $\frac{8^{5/2}}{8^{1/2}}$ **f.** $\frac{7^{2/3}}{7^{5/3}}$

3 EXPLORATION: Simplifying Products and Quotients of Radicals

Go to *BigIdeasMath.com* for an interactive tool to investigate this exploration.

Work with a partner. Use the properties of exponents to write each expression as a single radical. Then evaluate each expression. Use a calculator to check your answers.

a. $\sqrt{3} \bullet \sqrt{12}$ **b.** $\sqrt[3]{5} \bullet \sqrt[3]{25}$ **c.** $\sqrt[4]{27} \bullet \sqrt[4]{3}$

d. $\dfrac{\sqrt{98}}{\sqrt{2}}$ **e.** $\dfrac{\sqrt[4]{4}}{\sqrt[4]{1024}}$ **f.** $\dfrac{\sqrt[3]{625}}{\sqrt[3]{5}}$

Communicate Your Answer

4. How can you use properties of exponents to simplify products and quotients of radicals?

5. Simplify each expression.

a. $\sqrt{27} \bullet \sqrt{6}$ **b.** $\dfrac{\sqrt[3]{240}}{\sqrt[3]{15}}$ **c.** $\left(5^{1/2} \bullet 16^{1/4}\right)^2$

Name ____ Date __________

5.2 Notetaking with Vocabulary
For use after Lesson 5.2

In your own words, write the meaning of each vocabulary term.

simplest form of a radical

conjugate

like radicals

Core Concepts

Properties of Rational Exponents

Let a and b be real numbers and let m and n be rational numbers, such that the quantities in each property are real numbers.

Property Name	Definition	Example
Product of Powers	$a^m \bullet a^n = a^{m+n}$	$5^{1/2} \bullet 5^{3/2} = 5^{(1/2+3/2)} = 5^2 = 25$
Power of a Power	$\left(a^m\right)^n = a^{mn}$	$\left(3^{5/2}\right)^2 = 3^{(5/2 \bullet 2)} = 3^5 = 243$
Power of a Product	$(ab)^m = a^m b^m$	$(16 \bullet 9)^{1/2} = 16^{1/2} \bullet 9^{1/2} = 4 \bullet 3 = 12$
Negative Exponent	$a^{-m} = \frac{1}{a^m}, a \neq 0$	$36^{-1/2} = \frac{1}{36^{1/2}} = \frac{1}{6}$
Zero Exponent	$a^0 = 1, a \neq 0$	$213^0 = 1$
Quotient of Powers	$\frac{a^m}{a^n} = a^{m-n}, a \neq 0$	$\frac{4^{5/2}}{4^{1/2}} = 4^{(5/2-1/2)} = 4^2 = 16$
Power of a Quotient	$\left(\frac{a}{b}\right)^m = \frac{a^m}{b^m}, b \neq 0$	$\left(\frac{27}{64}\right)^{1/3} = \frac{27^{1/3}}{64^{1/3}} = \frac{3}{4}$

Notes:

Name______________________________ Date__________

5.2 Notetaking with Vocabulary (continued)

Properties of Radicals

Let a and b be real numbers and let n be an integer greater than 1.

Property Name	Definition	Example
Product Property	$\sqrt[n]{a \bullet b} = \sqrt[n]{a} \bullet \sqrt[n]{b}$	$\sqrt[3]{4} \bullet \sqrt[3]{2} = \sqrt[3]{8} = 2$
Quotient Property	$\sqrt[n]{\frac{a}{b}} = \frac{\sqrt[n]{a}}{\sqrt[n]{b}}, b \neq 0$	$\frac{\sqrt[4]{162}}{\sqrt[4]{2}} = \sqrt[4]{\frac{162}{2}} = \sqrt[4]{81} = 3$

Notes:

Extra Practice

In Exercises 1–4, use the properties of rational exponents to simplify the expression.

1. $\left(2^3 \bullet 3^3\right)^{-1/3}$

2. $\dfrac{10}{10^{-4/5}}$

3. $\left(\dfrac{52^5}{4^5}\right)^{1/6}$

4. $\dfrac{3^{1/3} \bullet 27^{2/3}}{8^{4/3}}$

5. Find simplified expressions for the perimeter and area of the given figure.

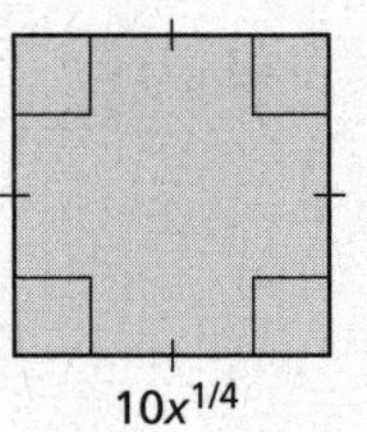

Name ____ Date __________

5.2 Notetaking with Vocabulary (continued)

In Exercises 6–8, use the properties of radicals to simplify the expression.

6. $\sqrt[6]{25} \bullet \sqrt[6]{625}$

7. $\dfrac{\sqrt{343}}{\sqrt{7}}$

8. $\dfrac{\sqrt[3]{25} \bullet \sqrt[3]{10}}{\sqrt[3]{2}}$

In Exercises 9–12, write the expression in simplest form.

9. $\sqrt[7]{384}$

10. $\sqrt[3]{\dfrac{5}{9}}$

11. $\dfrac{1}{4 - \sqrt{5}}$

12. $\dfrac{\sqrt{2}}{1 + \sqrt{6}}$

In Exercises 13–16, write the expression in simplest form. Assume all variables are positive.

13. $-2\sqrt[3]{5} + 40\sqrt[3]{5}$

14. $2(1250)^{1/4} - 5(32)^{1/4}$

15. $\dfrac{\sqrt[4]{x} \bullet \sqrt[4]{81x}}{\sqrt[4]{16x^{36}}}$

16. $\dfrac{21\left(x^{-3/2}\right)\left(\sqrt{y}\right)\left(z^{5/2}\right)}{7^{-1}\sqrt{x}\left(y^{-1/2}\right)z}$

Name____ Date__________

5.3 Graphing Radical Functions
For use with Exploration 5.3

Essential Question How can you identify the domain and range of a radical function?

1 EXPLORATION: Identifying Graphs of Radical Functions

Work with a partner. Match each function with its graph. Explain your reasoning. Then identify the domain and range of each function.

a. $f(x) = \sqrt{x}$

b. $f(x) = \sqrt[3]{x}$

c. $f(x) = \sqrt[4]{x}$

d. $f(x) = \sqrt[5]{x}$

A.

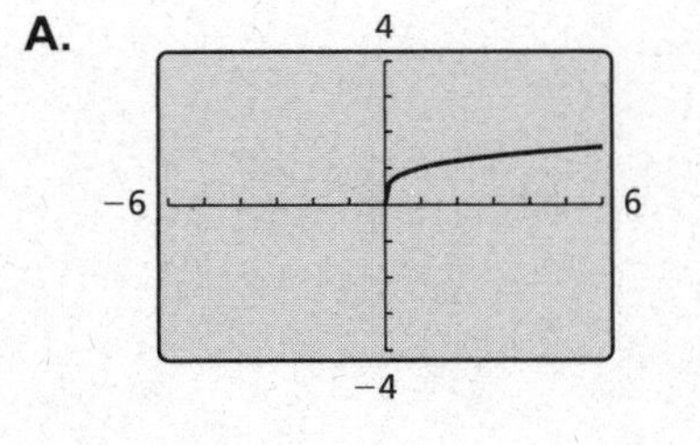

B.

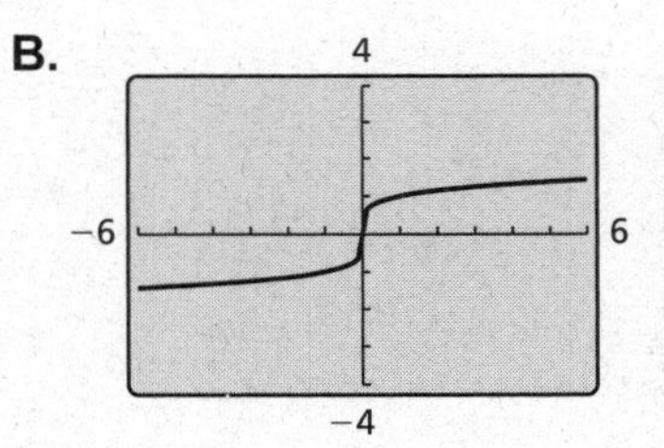

C.

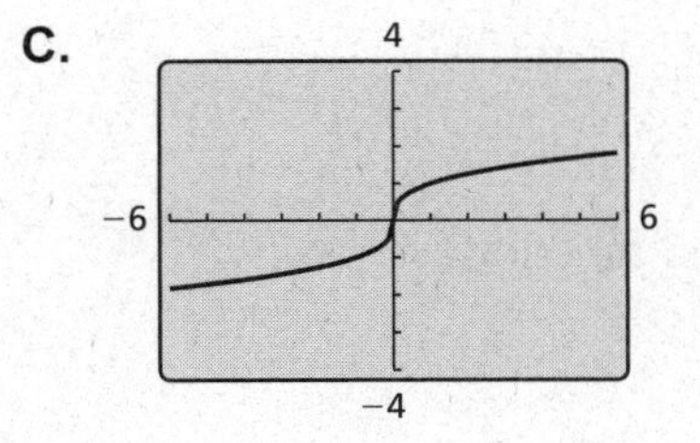

D.

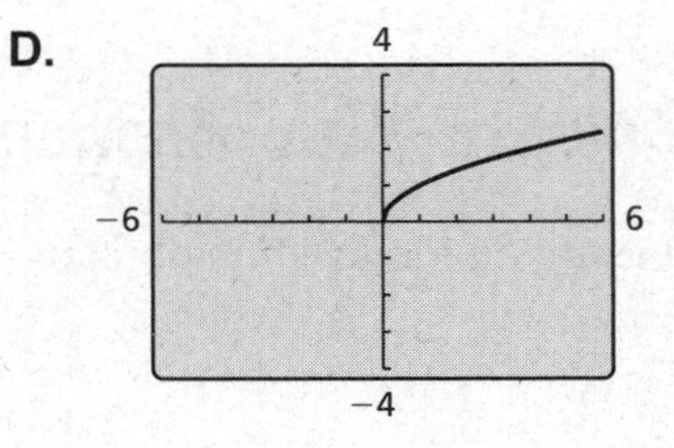

Name ______________________________ Date __________

2 EXPLORATION: Identifying Graphs of Transformations

Work with a partner. Match each transformation of $f(x) = \sqrt{x}$ with its graph. Explain your reasoning. Then identify the domain and range of each function.

a. $g(x) = \sqrt{x + 2}$

b. $g(x) = \sqrt{x - 2}$

c. $g(x) = \sqrt{x + 2} - 2$

d. $g(x) = -\sqrt{x + 2}$

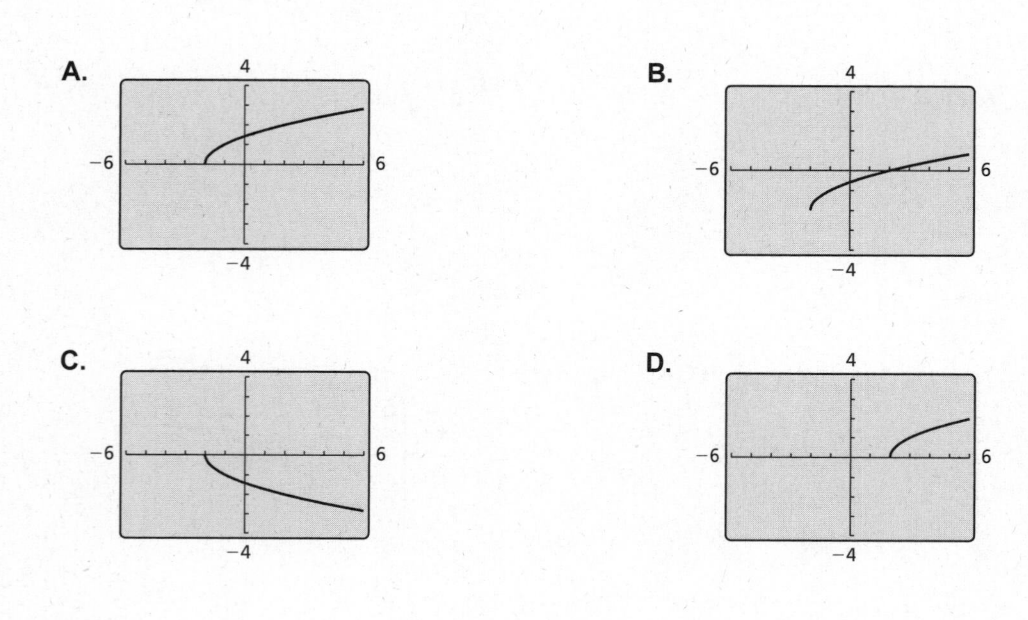

Communicate Your Answer

3. How can you identify the domain and range of a radical function?

4. Use the results of Exploration 1 to describe how the domain and range of a radical function are related to the index of the radical.

Name____ Date__________

5.3 Notetaking with Vocabulary
For use after Lesson 5.3

In your own words, write the meaning of each vocabulary term.

radical function

Core Concepts

Parent Functions for Square Root and Cube Root Functions

The parent function for the family of square root functions is $f(x) = \sqrt{x}$.

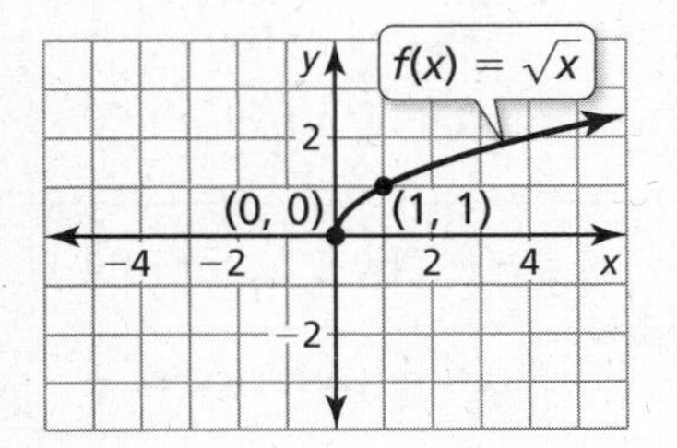

Domain: $x \geq 0$, Range: $y \geq 0$

The parent function for the family of cube root functions is $f(x) = \sqrt[3]{x}$.

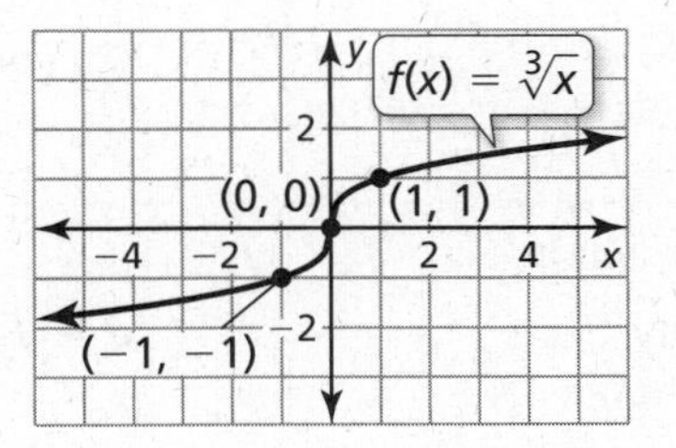

Domain and range: All real numbers

Notes:

Name ________________________ Date ________

Transformation	$f(x)$ Notation	Examples	
Horizontal Translation Graph shifts left or right.	$f(x - h)$	$g(x) = \sqrt{x - 2}$ $g(x) = \sqrt{x + 3}$	2 units right 3 units left
Vertical Translation Graph shifts up or down.	$f(x) + k$	$g(x) = \sqrt{x} + 7$ $g(x) = \sqrt{x} - 1$	7 units up 1 unit down
Reflection Graph flips over x- or y-axis.	$f(-x)$ $-f(x)$	$g(x) = \sqrt{-x}$ $g(x) = -\sqrt{x}$	in the y-axis in the x-axis
Horizontal Stretch or Shrink Graph stretches away from or shrinks toward y-axis.	$f(ax)$	$g(x) = \sqrt{3x}$ $g(x) = \sqrt{\frac{1}{2}x}$	shrink by a factor of $\frac{1}{3}$ stretch by a factor of 2
Vertical Stretch or Shrink Graph stretches away from or shrinks toward x-axis.	$a \bullet f(x)$	$g(x) = 4\sqrt{x}$ $g(x) = \frac{1}{5}\sqrt{x}$	stretch by a factor of 4 shrink by a factor of $\frac{1}{5}$

Notes:

Name____ Date__________

Extra Practice

In Exercises 1 and 2, graph the function. Identify the domain and range of each function.

1. $f(x) = \sqrt[3]{-3x} + 1$

2. $g(x) = 2(x - 5)^{1/2} - 4$

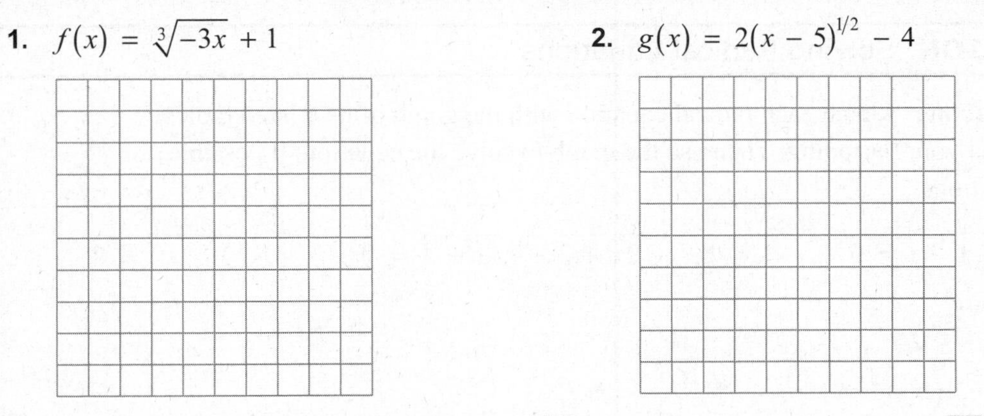

3. Describe the transformation of $f(x) = \sqrt[4]{2x} + 5$ represented by $g(x) = -\sqrt[4]{2x} - 5$.

4. Write a rule for g if g is a horizontal shrink by a factor of $\frac{5}{6}$, followed by a translation 10 units to the left of the graph of $f(x) = \sqrt[3]{15x + 1}$.

5. Use a graphing calculator to graph $8x = y^2 + 5$. Identify the vertex and the direction that the parabola opens.

6. Use a graphing calculator to graph $x^2 = 49 - y^2$. Identify the radius and the intercepts of the circle.

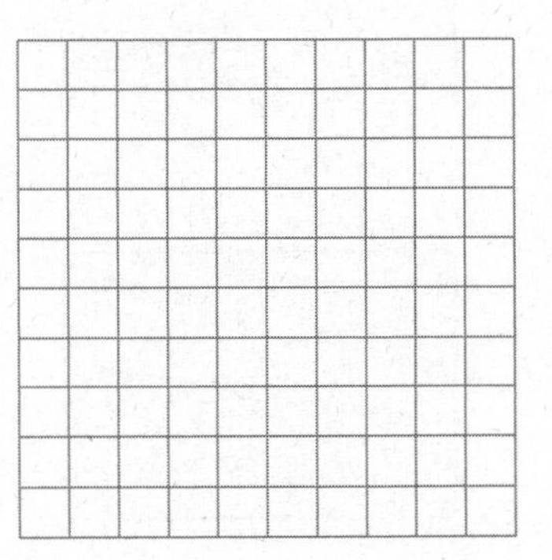

Name ______________________________ Date __________

5.4 Solving Radical Equations and Inequalities

For use with Exploration 5.4

Essential Question How can you solve a radical equation?

1 EXPLORATION: Solving Radical Equations

Work with a partner. Match each radical equation with the graph of its related radical function. Explain your reasoning. Then use the graph to solve the equation, if possible. Check your solutions.

a. $\sqrt{x-1}-1=0$ **b.** $\sqrt{2x+2}-\sqrt{x+4}=0$ **c.** $\sqrt{9-x^2}=0$

d. $\sqrt{x+2}-x=0$ **e.** $\sqrt{-x+2}-x=0$ **f.** $\sqrt{3x^2+1}=0$

A.

C.

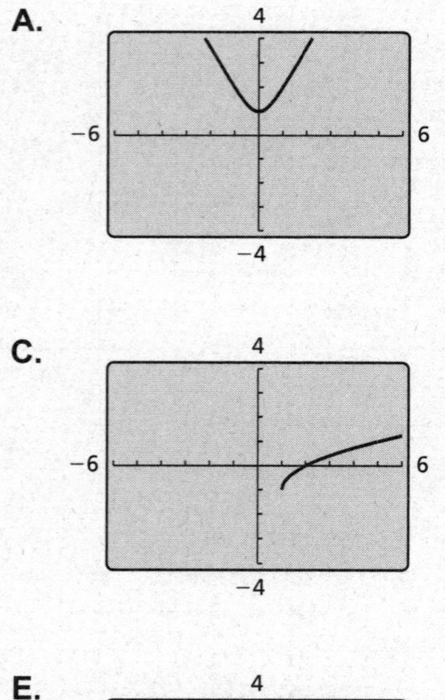

B.

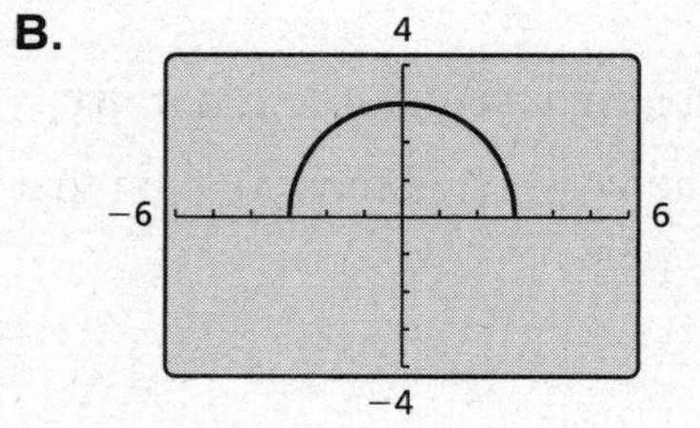

D.

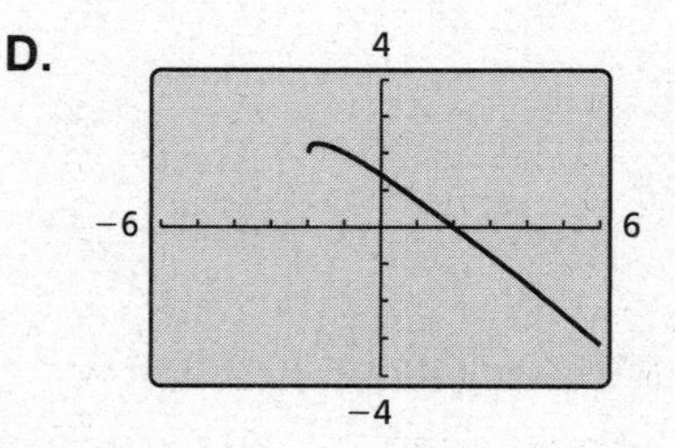

E.

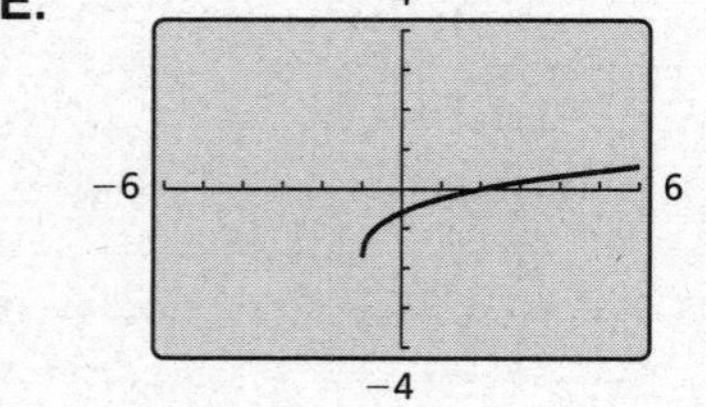

F.

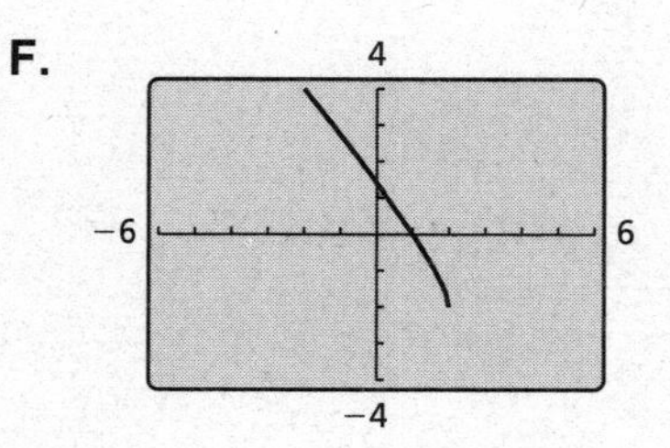

Name______________________________ Date__________

2 EXPLORATION: Solving Radical Equations

Go to *BigIdeasMath.com* for an interactive tool to investigate this exploration.

Work with a partner. Look back at the radical equations in Exploration 1. Suppose that you did not know how to solve the equations using a graphical approach.

a. Show how you could use a *numerical approach* to solve one of the equations. For instance, you might use a spreadsheet to create a table of values.

b. Show how you could use an *analytical approach* to solve one of the equations. For instance, look at the similarities between the equations in Exploration 1. What first step may be necessary so you could square each side to eliminate the radical(s)? How would you proceed to find the solution?

Communicate Your Answer

3. How can you solve a radical equation?

4. Would you prefer to use a graphical, numerical, or analytical approach to solve the given equation? Explain your reasoning. Then solve the equation.

$$\sqrt{x+3} - \sqrt{x-2} = 1$$

Name ______________________________ Date __________

5.4 Notetaking with Vocabulary

For use after Lesson 5.4

In your own words, write the meaning of each vocabulary term.

radical equation

extraneous solutions

Core Concepts

Solving Radical Equations

To solve a radical equation, follow these steps:

Step 1 Isolate the radical on one side of the equation, if necessary.

Step 2 Raise each side of the equation to the same exponent to eliminate the radical and obtain a linear, quadratic, or other polynomial equation.

Step 3 Solve the resulting equation using techniques you learned in previous chapters. Check your solution.

Notes:

Name____ Date__________

Extra Practice

In Exercises 1–10, solve the equation. Check your solution(s).

1. $\sqrt{1 - x} = 7$

2. $\sqrt[3]{5x + 1} = -4$

3. $\frac{1}{4}\sqrt[4]{2x} + 6 = 10$

4. $2\sqrt[3]{13x - 5} = 10$

5. $x - 7 = \sqrt{x - 5}$

6. $\sqrt[3]{486 - 27x^3} = 3x$

7. $4\sqrt{x + 1} = x + 1$

8. $\sqrt{2x + 2} - 3\sqrt{x + 1} = 0$

9. $2 - \sqrt[4]{2x - 6} = 14$

10. $\sqrt{x + 7} + 2 = \sqrt{3 - x}$

5.4 Notetaking with Vocabulary (continued)

In Exercises 11 and 12, solve the equation. Check your solution(s).

11. $\frac{1}{2}x^{5/2} = 16$

12. $(6x + 10)^{7/3} + 28 = 156$

In Exercises 13–15, solve the inequality.

13. $-4\sqrt{x - 1} + 3 \geq -1$

14. $\sqrt[3]{\frac{2}{3}x + 1} < 6$

15. $2\sqrt{\frac{3}{4}x} - 39 \leq -25$

16. In basketball, the term "hang time" is the amount of time that a player is suspended in the air when making a basket. To win a slam-dunk contest, players try to maximize their hang time. A player's hang time is given by the equation $t = 0.5\sqrt{h}$, where t is the time (in seconds) and h is the height (in feet) of the jump. The second-place finisher of a slam-dunk contest had a hang time of 1 second, and the winner had a hang time of 1.2 seconds. How many feet higher did the winner jump than the second-place finisher?

Name____ Date__________

5.5 Performing Function Operations

For use with Exploration 5.5

Essential Question How can you use the graphs of two functions to sketch the graph of an arithmetic combination of the two functions?

Just as two real numbers can be combined by the operations of addition, subtraction, multiplication, and division to form other real numbers, two functions can be combined to form other functions. For example, the functions $f(x) = 2x - 3$ and $g(x) = x^2 - 1$ can be combined to form the sum, difference, product, or quotient of f and g.

$f(x) + g(x) = (2x - 3) + (x^2 - 1) = x^2 + 2x - 4$ sum

$f(x) - g(x) = (2x - 3) - (x^2 - 1) = -x^2 + 2x - 2$ difference

$f(x) \bullet g(x) = (2x - 3)(x^2 - 1) = 2x^3 - 3x^2 - 2x + 3$ product

$\dfrac{f(x)}{g(x)} = \dfrac{2x - 3}{x^2 - 1}, x \neq \pm 1$ quotient

1 EXPLORATION: Graphing the Sum of Two Functions

Go to *BigIdeasMath.com* for an interactive tool to investigate this exploration.

Work with a partner. Use the graphs of f and g to sketch the graph of $f + g$. Explain your steps.

Sample Use a compass or a ruler to measure the distance from a point on the graph of g to the x-axis. Then add this distance to the point with the same x-coordinate on the graph of f. Plot the new point. Repeat this process for several points. Finally, draw a smooth curve through the new points to obtain the graph of $f + g$.

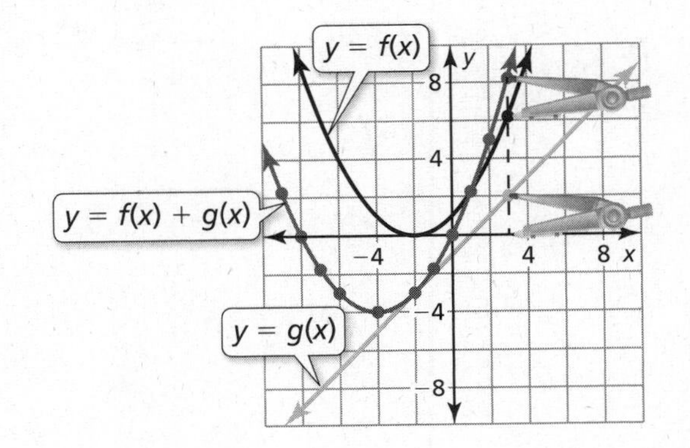

1 EXPLORATION: Graphing the Sum of Two Functions (continued)

a.

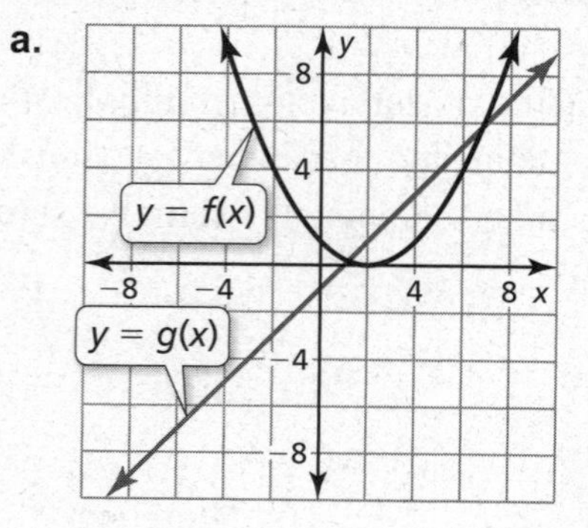

b.

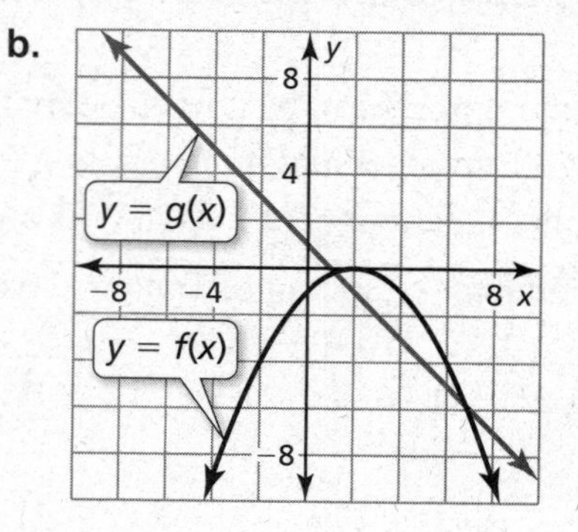

Communicate Your Answer

2. How can you use the graphs of two functions to sketch the graph of an arithmetic combination of the two functions?

3. Check your answers in Exploration 1 by writing equations for f and g, adding the functions, and graphing the sum.

Name______________________________ Date__________

5.5 Notetaking with Vocabulary

For use after Lesson 5.5

In your own words, write the meaning of each vocabulary term.

domain

scientific notation

Core Concepts

Operations on Functions

Let f and g be any two functions. A new function can be defined by performing any of the four basic operations on f and g.

Operation	Definition	Example: $f(x) = 5x$, $g(x) = x + 2$
Addition	$(f + g)(x) = f(x) + g(x)$	$(f + g)(x) = 5x + (x + 2) = 6x + 2$
Subtraction	$(f - g)(x) = f(x) - g(x)$	$(f - g)(x) = 5x - (x + 2) = 4x - 2$
Multiplication	$(fg)(x) = f(x) \bullet g(x)$	$(fg)(x) = 5x(x + 2) = 5x^2 + 10x$
Division	$\left(\frac{f}{g}\right)(x) = \frac{f(x)}{g(x)}$	$\left(\frac{f}{g}\right)(x) = \frac{5x}{x + 2}$

The domains of the sum, difference, product, and quotient functions consist of the x-values that are in the domains of both f and g. Additionally, the domain of the quotient does not include x-values for which $g(x) = 0$.

Notes:

Name ____ Date __________

Extra Practice

In Exercises 1–4, find $(f + g)(x)$ and $(f - g)(x)$ and state the domain of each. Then evaluate $f + g$ and $f - g$ for the given value of x.

1. $f(x) = -\frac{1}{2}\sqrt[3]{x}, g(x) = \frac{9}{2}\sqrt[3]{x}; x = -1000$

2. $f(x) = -x^2 - 3x + 8, g(x) = 6x - 3x^2; x = -1$

3. $f(x) = 4x^3 + 12, g(x) = 2x^2 - 3x^3 + 9; x = 2$

4. $f(x) = 5\sqrt[4]{x} + 1, g(x) = -3\sqrt[4]{x} - 2; x = 1$

In Exercises 5–8, find $(fg)(x)$ and $\left(\frac{f}{g}\right)(x)$ and state the domain of each. Then evaluate fg and $\frac{f}{g}$ for the given value of x.

5. $f(x) = -x^3, g(x) = 2\sqrt[3]{x}; x = -64$

6. $f(x) = 12x, g(x) = 11x^{1/2}; x = 4$

7. $f(x) = 0.25x^{1/3}, g(x) = -4x^{3/2}; x = 1$

8. $f(x) = 36x^{7/4}, g(x) = 4x^{1/2}; x = 16$

9. The graphs of the functions $f(x) = x^2 - 4x + 4$ and $g(x) = 4x - 5$ are shown. Which graph represents the function $f + g$? the function $f - g$? Explain your reasoning.

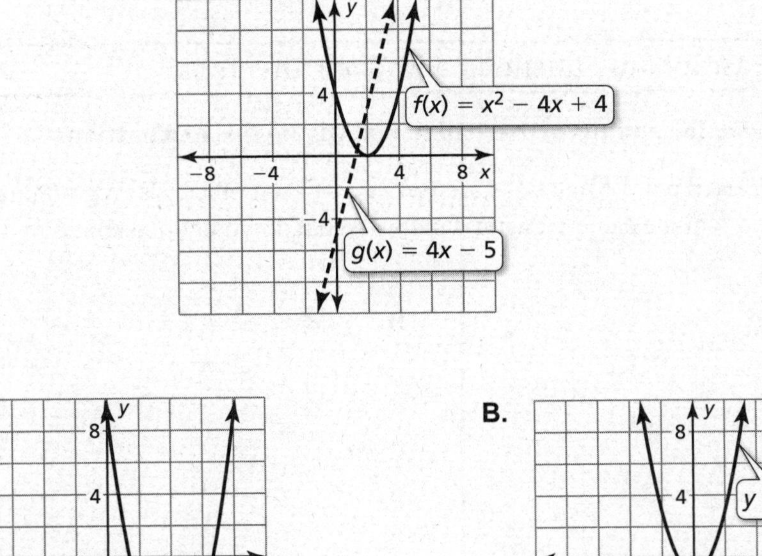

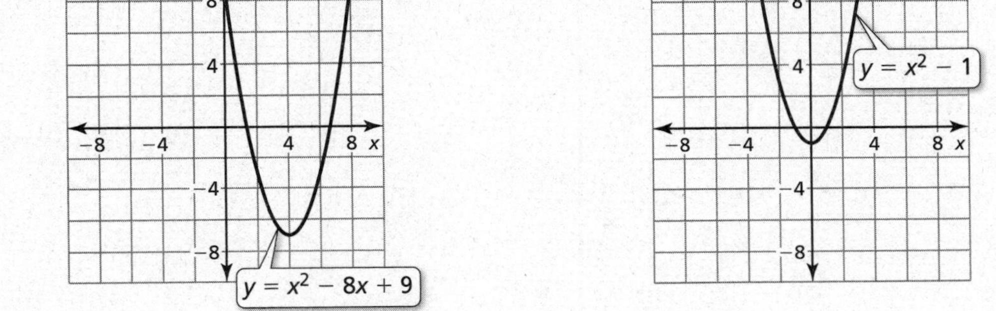

10. The variable x represents the number of pages of a textbook to be printed. The cost C to print the textbook can be modeled by the equation $C(x) = 0.2x^2 + 10$. The selling price P of the textbook can be modeled by the equation $P(x) = 0.05x^2 + 20$.

a. Find $(P - C)(x)$.

b. Explain what $(P - C)(x)$ represents.

Name ______________________________ Date __________

5.6 Inverse of a Function

For use with Exploration 5.6

Essential Question How can you sketch the graph of the inverse of a function?

1 EXPLORATION: Graphing Functions and Their Inverses

Go to *BigIdeasMath.com* for an interactive tool to investigate this exploration.

Work with a partner. Each pair of functions are *inverses* of each other. Use a graphing calculator to graph f and g in the same viewing window. What do you notice about the graphs?

a. $f(x) = 4x + 3$

$g(x) = \dfrac{x - 3}{4}$

b. $f(x) = x^3 + 1$

$g(x) = \sqrt[3]{x - 1}$

c. $f(x) = \sqrt{x - 3}$

$g(x) = x^2 + 3, x \geq 0$

d. $f(x) = \dfrac{4x + 4}{x + 5}$

$g(x) = \dfrac{4 - 5x}{x - 4}$

2 EXPLORATION: Sketching Graphs of Inverse Functions

Go to *BigIdeasMath.com* for an interactive tool to investigate this exploration.

Work with a partner. Use the graph of f to sketch the graph of g, the inverse function of f, on the same set of coordinate axes. Explain your reasoning.

a.

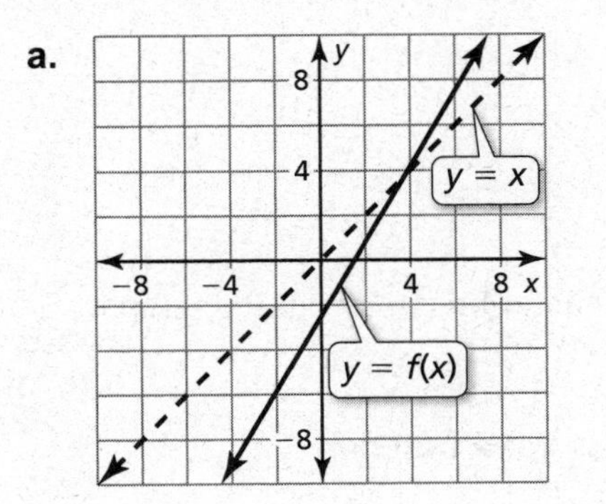

b.

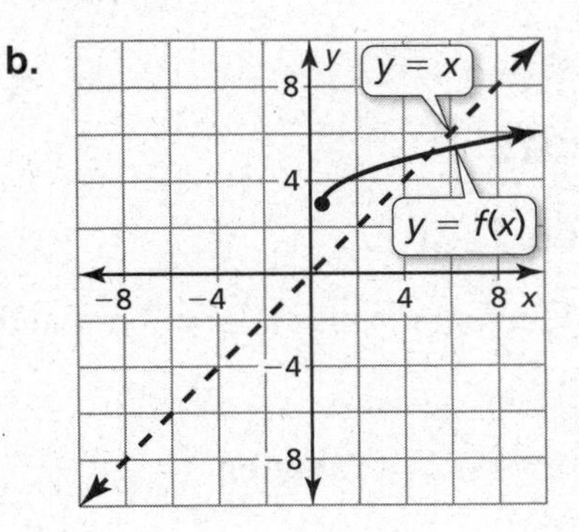

c.

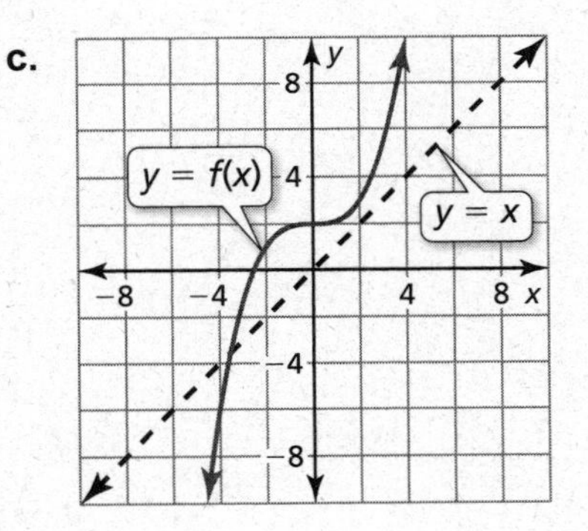

d.

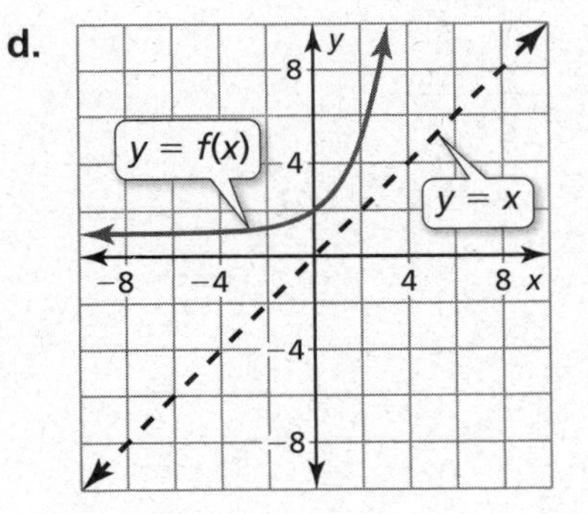

Communicate Your Answer

3. How can you sketch the graph of the inverse of a function?

4. In Exploration 1, what do you notice about the relationship between the equations of f and g? Use your answer to find g, the inverse function of

$$f(x) = 2x - 3.$$

Use a graph to check your answer.

Name ______________________________ Date ________

5.6 Notetaking with Vocabulary
For use after Lesson 5.6

In your own words, write the meaning of each vocabulary term.

inverse functions

Core Concepts

Horizontal Line Test

The inverse of a function f is also a function if and only if no horizontal line intersects the graph of f more than once.

Inverse is a function

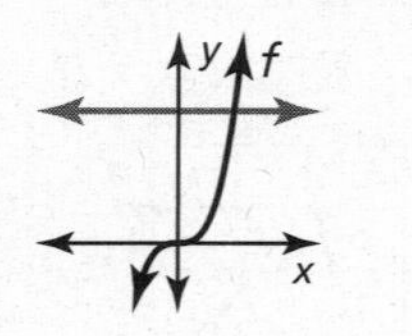

Inverse is not a function

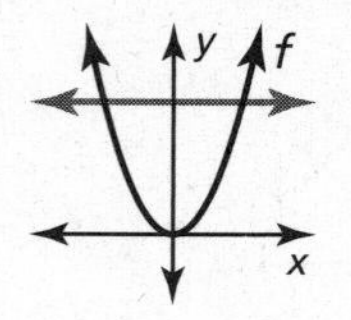

Notes:

Name____ Date__________

Extra Practice

In Exercises 1–3, solve $y = f(x)$ for x. Then find the input(s) when the output is −6.

1. $f(x) = 2x - 1$

2. $f(x) = 1 - x^2$

3. $f(x) = (x - 1)^3 + 2$

In Exercises 4–6, find the inverse of the function. Then graph the function and its inverse.

4. $f(x) = 10x$

5. $f(x) = -\frac{1}{5}x - 7$

6. $f(x) = \frac{3}{4}x + \frac{5}{8}$

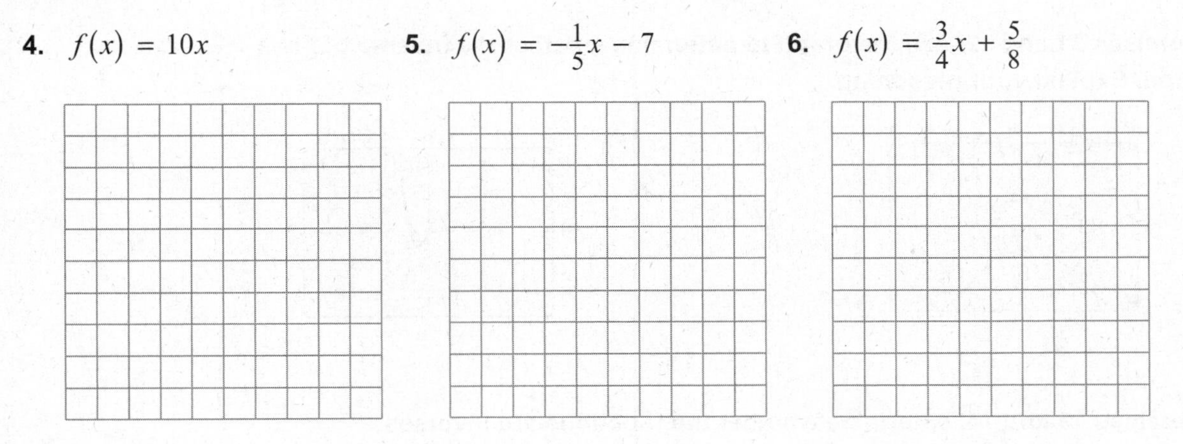

In Exercises 7 and 8, determine whether each pair of functions *f* and *g* are inverses. Explain your reasoning.

7.

x	−4	−3	−2	−1	0	1
f(x)	17	13	9	5	1	−3

x	17	13	9	5	1	−3
g(x)	−4	−3	−2	−1	0	−1

8.

x	1	2	3	4	5	6
f(x)	−1	−2	−4	−5	−8	−10

x	−1	−0.5	−0.25	−0.2	−0.125	−0.1
g(x)	1	2	3	4	5	6

Name ______________________________ Date __________

 Notetaking with Vocabulary (continued)

In Exercises 9 and 10, find the inverse of the function. Then graph the function and its inverse.

9. $f(x) = (x+2)^3$

10. $f(x) = \frac{1}{3}x^4, x \geq 0$

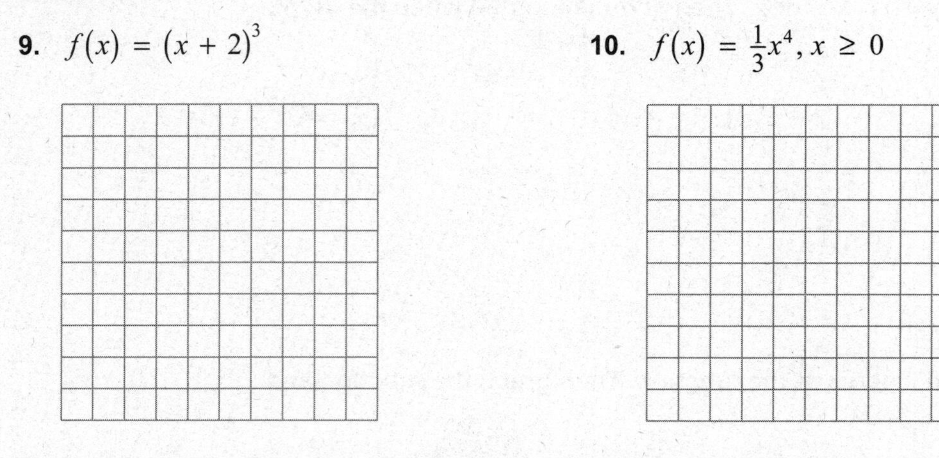

In Exercises 11 and 12, use the graph to determine whether the inverse of *f* is a function. Explain your reasoning.

11.

12.

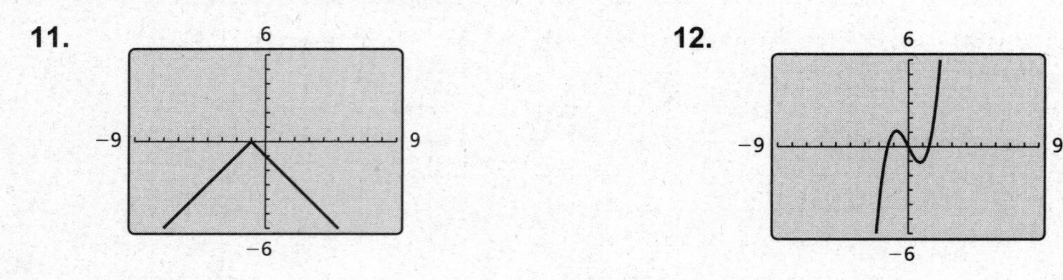

In Exercises 13 and 14, determine whether the functions are inverses.

13. $f(x) = \dfrac{4x}{5} - 1, g(x) = \dfrac{5x+1}{4}$

14. $f(x) = -(x-2)^5 + 6, g(x) = 2 + (6-x)^{1/5}$

Name____ Date__________

Chapter 6 Maintaining Mathematical Proficiency

Evaluate the expression.

1. $-4 \bullet 5^3$ **2.** $(-3)^4$ **3.** $-\left(\frac{7}{8}\right)^2$ **4.** $\left(\frac{3}{10}\right)^3$

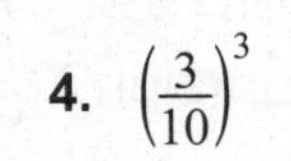

Find the domain and range of the function represented by the graph.

5.

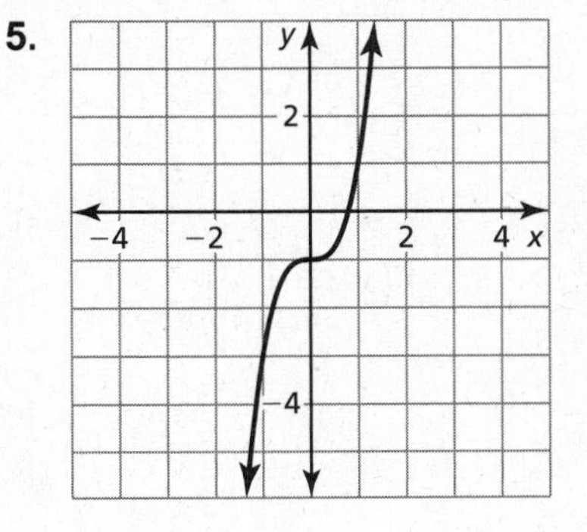

6.

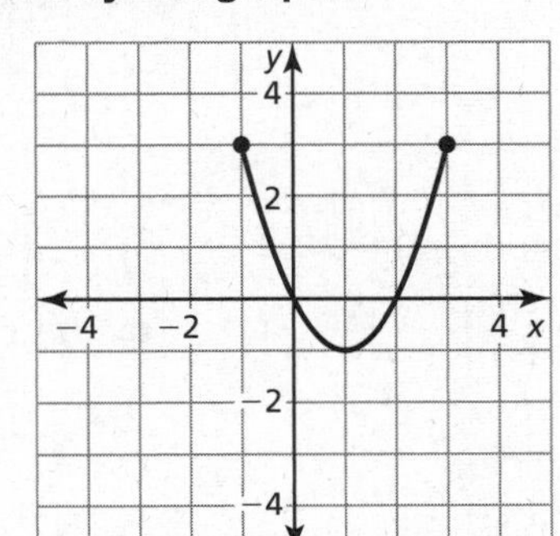

7.

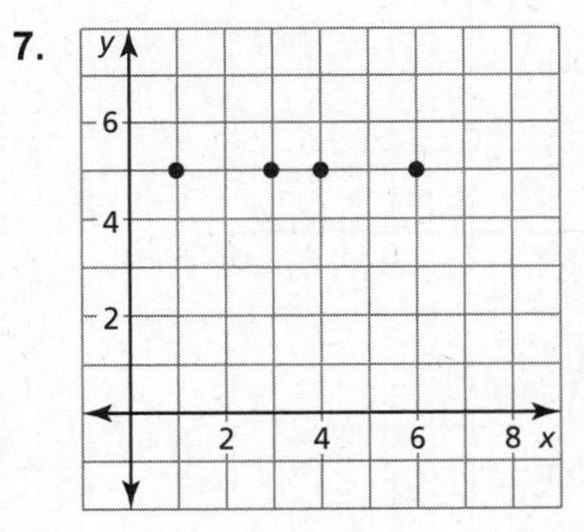

8.

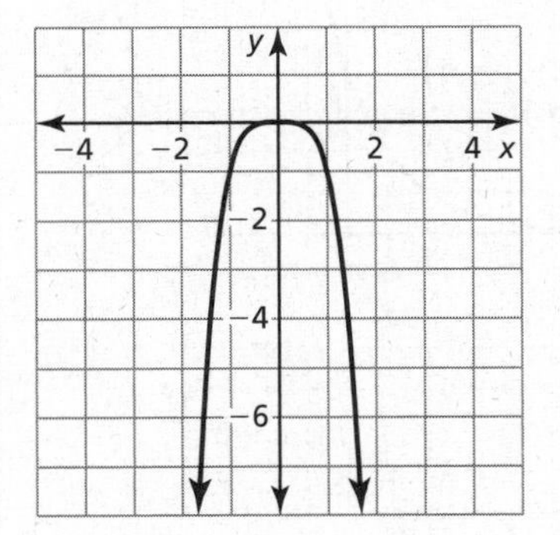

9. Is the expression "the sum of the square of x and the square of the opposite of x" equivalent to 0 or $2x^2$? Explain your reasoning.

Name ______________________________ Date ________

6.1 Exponential Growth and Decay Functions

For use with Exploration 6.1

Essential Question What are some of the characteristics of the graph of an exponential function?

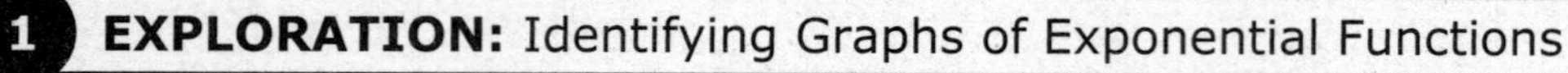

1 EXPLORATION: Identifying Graphs of Exponential Functions

Work with a partner. Match each exponential function with its graph. Use a table of values to sketch the graph of the function, if necessary.

a. $f(x) = 2^x$ **b.** $f(x) = 3^x$ **c.** $f(x) = 4^x$

d. $f(x) = \left(\frac{1}{2}\right)^x$ **e.** $f(x) = \left(\frac{1}{3}\right)^x$ **f.** $f(x) = \left(\frac{1}{4}\right)^x$

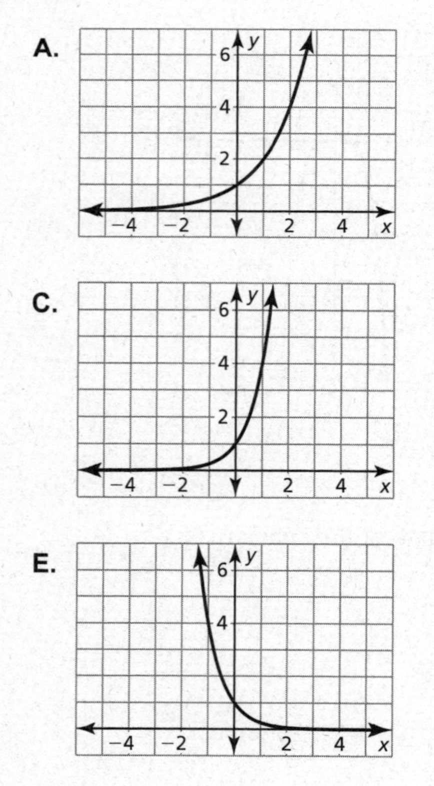

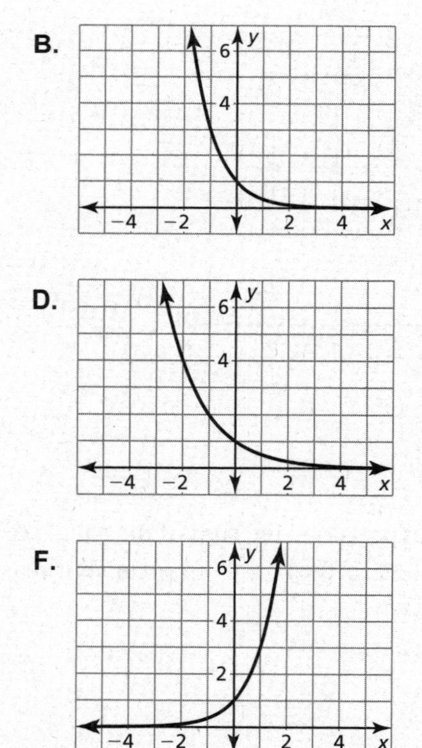

2 EXPLORATION: Characteristics of Graphs of Exponential Functions

Work with a partner. Use the graphs in Exploration 1 to determine the domain, range, and y-intercept of the graph of $f(x) = b^x$, where b is a positive real number other than 1. Explain your reasoning.

Communicate Your Answer

3. What are some of the characteristics of the graph of an exponential function?

4. In Exploration 2, is it possible for the graph of $f(x) = b^x$ to have an x-intercept? Explain your reasoning.

Name ______________________________ Date __________

6.1 Notetaking with Vocabulary

For use after Lesson 6.1

In your own words, write the meaning of each vocabulary term.

exponential function

exponential growth function

growth factor

asymptote

exponential decay function

decay factor

Core Concepts

Parent Function for Exponential Growth Functions

The function $f(x) = b^x$, where $b > 1$, is the parent function for the family of exponential growth functions with base b. The graph shows the general shape of an exponential growth function.

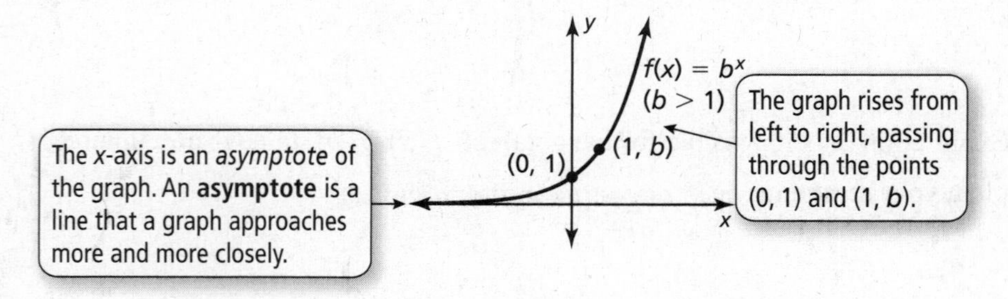

The domain of $f(x) = b^x$ is all real numbers. The range is $y > 0$.

Notes:

Parent Function for Exponential Decay Functions

The function $f(x) = b^x$, where $0 < b < 1$, is the parent function for the family of exponential decay functions with base b. The graph shows the general shape of an exponential decay function.

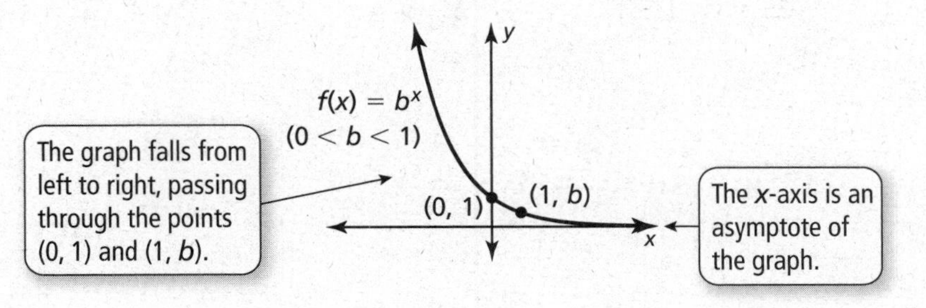

The domain of $f(x) = b^x$ is all real numbers. The range is $y > 0$.

Notes:

Compound Interest

Consider an initial principal P deposited in an account that pays interest at an annual rate r (expressed as a decimal), compounded n times per year. The amount A in the account after t years is given by

$$A = P\left(1 + \frac{r}{n}\right)^{nt}.$$

Notes:

Name ______________________________ Date __________

Extra Practice

In Exercises 1–4, tell whether the function represents *exponential growth* or *exponential decay.* Then graph the function.

1. $y = \left(\frac{1}{12}\right)^x$ **2.** $y = (1.5)^x$ **3.** $y = \left(\frac{7}{2}\right)^x$ **4.** $y = (0.8)^x$

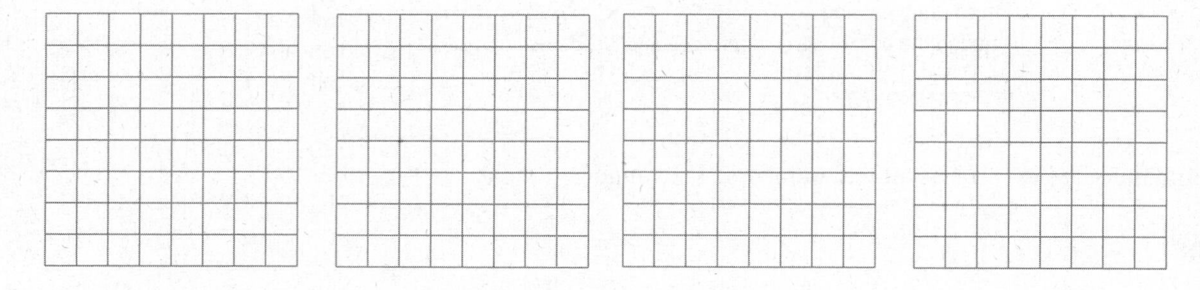

5. The number of bacteria y (in thousands) in a culture can be approximated by the model $y = 100(1.99)^t$, where t is the number of hours the culture is incubated.

a. Tell whether the model represents exponential growth or exponential decay.

b. Identify the hourly percent increase or decrease in the number of bacteria.

c. Estimate when the number of bacteria will be 1,000,000.

In Exercises 6 and 7, use the given information to find the amount *A* in the account earning compound interest after 5 years when the principal is $1250.

6. $r = 2.25\%$, compounded quarterly

7. $r = 1.25\%$, compounded daily

Name______________________________ Date__________

The Natural Base e

For use with Exploration 6.2

Essential Question What is the natural base *e*?

So far in your study of mathematics, you have worked with special numbers such as π and i. Another special number is called the *natural base* and is denoted by e. The natural base e is irrational, so you cannot find its exact value.

1 EXPLORATION: Approximating the Natural Base *e*

Go to *BigIdeasMath.com* for an interactive tool to investigate this exploration.

Work with a partner. One way to approximate the natural base e is to approximate the sum

$$1 + \frac{1}{1} + \frac{1}{1 \bullet 2} + \frac{1}{1 \bullet 2 \bullet 3} + \frac{1}{1 \bullet 2 \bullet 3 \bullet 4} + \cdots.$$

Use a spreadsheet or a graphing calculator to approximate this sum. Explain the steps you used. How many decimal places did you use in your approximation?

2 EXPLORATION: Approximating the Natural Base *e*

Work with a partner. Another way to approximate the natural base e is to consider the expression

$$\left(1 + \frac{1}{x}\right)^x.$$

As x increases, the value of this expression approaches the value of e. Complete the table. Then use the results in the table to approximate e. Compare this approximation to the one you obtained in Exploration 1.

x	10^1	10^2	10^3	10^4	10^5	10^6
$\left(1 + \frac{1}{x}\right)^x$						

Name ______________________________ Date __________

3 EXPLORATION: Graphing a Natural Base Function

Go to *BigIdeasMath.com* for an interactive tool to investigate this exploration.

Work with a partner. Use your approximate value of e in Exploration 1 or 2 to complete the table. Then sketch the graph of the *natural base exponential function* $y = e^x$. You can use a graphing calculator and the e^x key to check your graph. What are the domain and range of $y = e^x$? Justify your answers.

x	-2	-1	0	1	2
$y = e^x$					

Communicate Your Answer

4. What is the natural base e?

5. Repeat Exploration 3 for the natural base exponential function $y = e^{-x}$. Then compare the graph of $y = e^x$ to the graph of $y = e^{-x}$.

x	-2	-1	0	1	2
$y = e^{-x}$					

6. The natural base e is used in a wide variety of real-life applications. Use the Internet or some other reference to research some of the real-life applications of e.

Name____ Date__________

6.2 Notetaking with Vocabulary
For use after Lesson 6.2

In your own words, write the meaning of each vocabulary term.

natural base *e*

Core Concepts

The Natural Base e

The natural base e is irrational. It is defined as follows:

As x approaches $+\infty$, $\left(1 + \frac{1}{x}\right)^x$ approaches $e \approx 2.71828182846$.

Notes:

Natural Base Functions

A function of the form $y = ae^{rx}$ is called a *natural base exponential function*.

- When $a > 0$ and $r > 0$, the function is an exponential growth function.
- When $a > 0$ and $r < 0$, the function is an exponential decay function.

The graphs of the basic functions $y = e^x$ and $y = e^{-x}$ are shown.

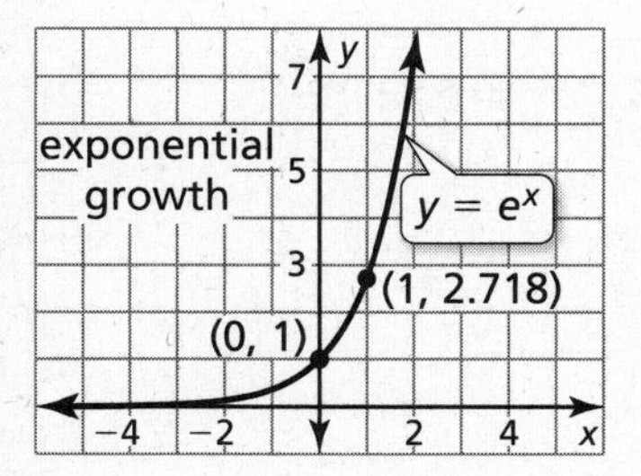

exponential decay
$y = e^{-x}$
(0, 1)
(1, 0.368)

Notes:

Name ______________________________ Date __________

6.2 Notetaking with Vocabulary (continued)

Continuously Compounded Interest

When interest is compounded *continuously*, the amount A in an account after t years is given by the formula

$$A = Pe^{rt}$$

where P is the principal and r is the annual interest rate expressed as a decimal.

Notes:

Extra Practice

In Exercises 1–4, simplify the expression.

1. $e^{-9} \bullet e^{12}$ **2.** $\dfrac{25e^2}{35e^7}$ **3.** $\left(2e^{-3x}\right)^5 \bullet 2e^{x+1}$ **4.** $\sqrt[4]{16e^{24x}}$

In Exercises 5–8, tell whether the function represents *exponential growth* or *exponential decay*. Then graph the function.

5. $y = 2e^{-x}$ **6.** $y = 0.75e^{4x}$ **7.** $y = 5e^{0.25x}$ **8.** $y = 0.8e^{-3x}$

In Exercises 9–11, use a table of values or a graphing calculator to graph the function. Then identify the domain and range.

9. $y = e^x - 4$

10. $y = 2e^{x+3}$

11. $y = -e^x + 5$

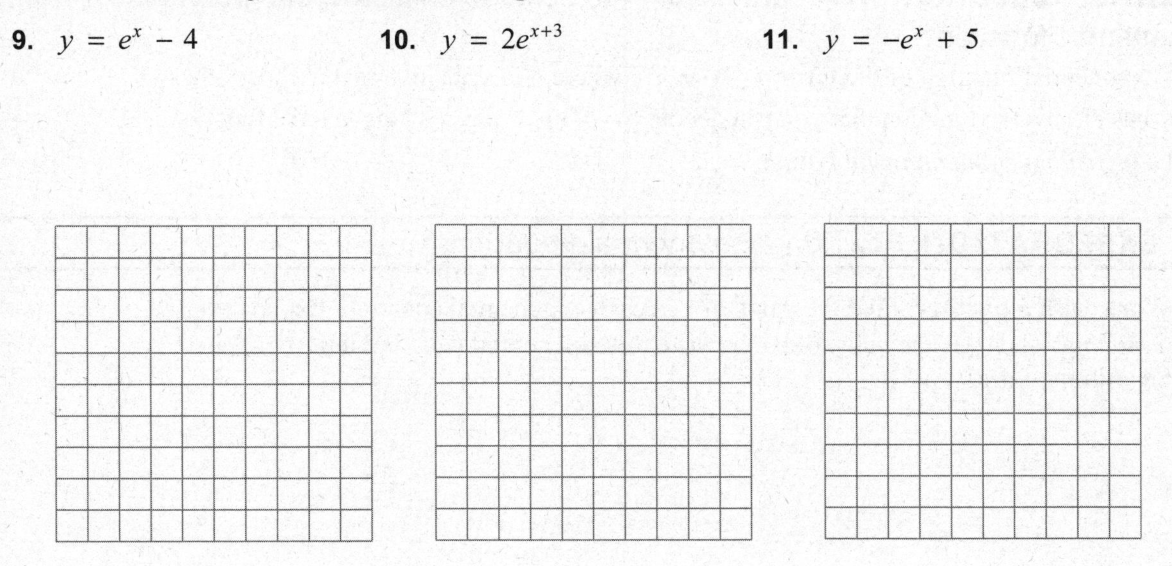

12. The population of Evans City is currently 48,500 and is declining at a rate of 2.5% each year. You can model the population of Evans City by the equation $P_t = P_c e^{rt}$, where P_c is the current population, P_t is the population after t years, and r is the decimal rate of decline per year. Predict the population of Evans City after 25 years.

13. Your parents will need $25,000 in 10 years to pay for your brother's college tuition. They can invest in an account with an interest rate of 9.8% that compounds continuously. How much should your parents invest today in order to have your brother's full tuition available in 10 years?

Name ______________________ Date __________

6.3 Logarithms and Logarithmic Functions

For use with Exploration 6.3

Essential Question What are some of the characteristics of the graph of a logarithmic function?

Every exponential function of the form $f(x) = b^x$, where b is a positive real number other than 1, has an inverse function that you can denote by $g(x) = \log_b x$. This inverse function is called a *logarithmic function with base b.*

1 EXPLORATION: Rewriting Exponential Equations

Work with a partner. Find the value of x in each exponential equation. Explain your reasoning. Then use the value of x to rewrite the exponential equation in its equivalent logarithmic form, $x = \log_b y$.

a. $2^x = 8$

b. $3^x = 9$

c. $4^x = 2$

d. $5^x = 1$

e. $5^x = \frac{1}{5}$

f. $8^x = 4$

2 EXPLORATION: Graphing Exponential and Logarithmic Functions

Go to *BigIdeasMath.com* for an interactive tool to investigate this exploration.

Work with a partner. Complete each table for the given exponential function. Use the results to complete the table for the given logarithmic function. Explain your reasoning. Then sketch the graphs of f and g in the same coordinate plane.

a.

x	−2	−1	0	1	2
$f(x) = 2^x$					

x					
$g(x) = \log_2 x$	−2	−1	0	1	2

Name____ Date__________

6.3 Logarithms and Logarithmic Functions (continued)

2 EXPLORATION: Graphing Exponential and Logarithmic Functions (continued)

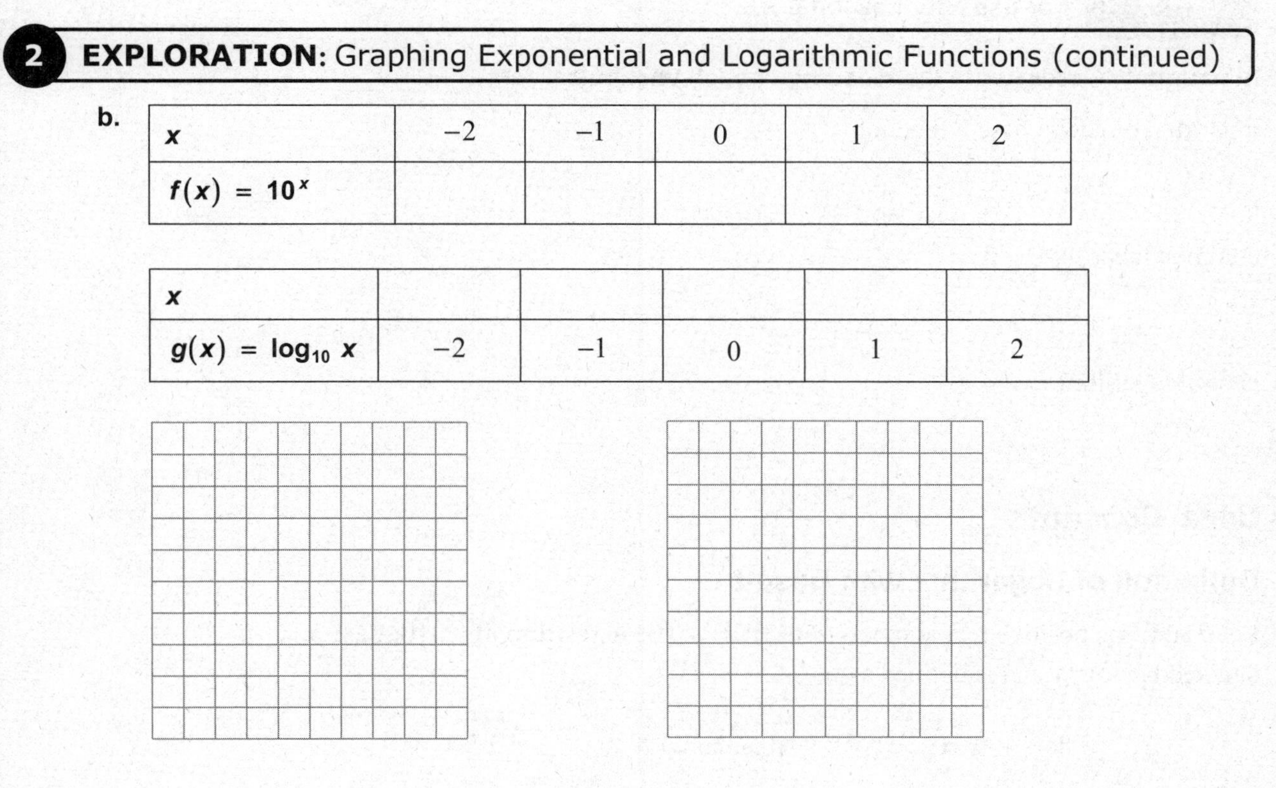

b.

x	-2	-1	0	1	2
$f(x) = 10^x$					

x					
$g(x) = \log_{10} x$	-2	-1	0	1	2

3 EXPLORATION: Characteristics of Graphs of Logarithmic Functions

Work with a partner. Use the graphs you sketched in Exploration 2 to determine the domain, range, x-intercept, and asymptote of the graph of $g(x) = \log_b x$, where b is a positive real number other than 1. Explain your reasoning.

Communicate Your Answer

4. What are some of the characteristics of the graph of a logarithmic function?

5. How can you use the graph of an exponential function to obtain the graph of a logarithmic function?

Name ______________________________ Date __________

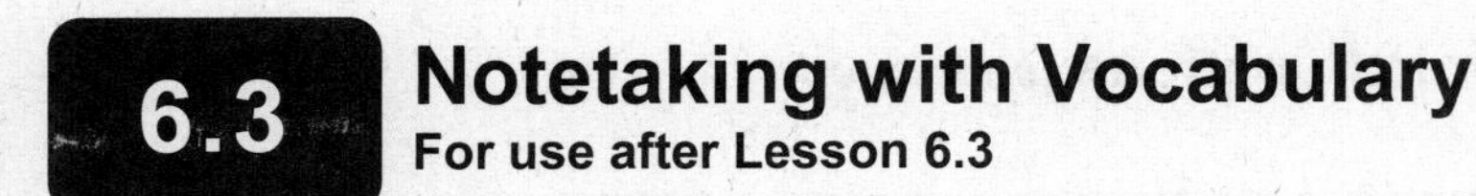

6.3 Notetaking with Vocabulary

For use after Lesson 6.3

In your own words, write the meaning of each vocabulary term.

logarithm of y with base b function

common logarithm

natural logarithm

Core Concepts

Definition of Logarithm with Base b

Let b and y be positive real numbers with $b \neq 1$. The **logarithm of y with base b** is denoted by $\log_b y$ and is defined as

$$\log_b y = x \quad \text{if and only if} \quad b^x = y.$$

The expression $\log_b y$ is read as "log base b of y."

Notes:

Name____ Date__________

Parent Graphs for Logarithmic Functions

The graph of $f(x) = \log_b x$ is shown below for $b > 1$ and for $0 < b < 1$. Because $f(x) = \log_b x$ and $g(x) = b^x$ are inverse functions, the graph of $f(x) = \log_b x$ is the reflection of the graph of $g(x) = b^x$ in the line $y = x$.

Graph of $f(x) = \log_b x$ for $b > 1$ **Graph of $f(x) = \log_b x$ for $0 < b < 1$**

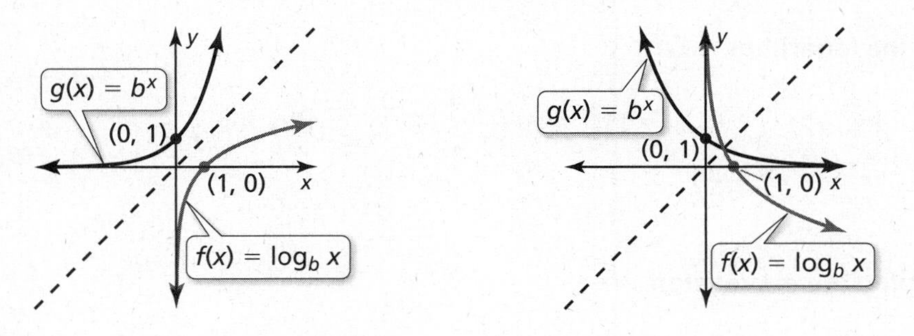

Note that the y-axis is a vertical asymptote of the graph of $f(x) = \log_b x$. The domain of $f(x) = \log_b x$ is $x > 0$, and the range is all real numbers.

Notes:

Extra Practice

In Exercises 1–4, rewrite the equation in exponential form.

1. $\log_{10} 1000 = 3$ **2.** $\log_5 \frac{1}{25} = -2$ **3.** $\log_{10} 1 = 0$ **4.** $\log_{1/4} 64 = -3$

Name ______________________________ Date __________

6.3 Notetaking with Vocabulary (continued)

In Exercises 5–8, rewrite the equation in logarithmic form.

5. $12^2 = 144$ **6.** $20^{-1} = \frac{1}{20}$ **7.** $216^{1/3} = 6$ **8.** $4^0 = 1$

In Exercises 9–12, evaluate the logarithm.

9. $\log_4 64$ **10.** $\log_{1/8} 1$ **11.** $\log_2 \frac{1}{32}$ **12.** $\log_{1/25} \frac{1}{5}$

In Exercises 13 and 14, simplify the expression.

13. $13^{\log_{13} 6}$ **14.** $\ln e^{x^3}$

In Exercises 15 and 16, find the inverse of the function.

15. $y = 15^x + 10$ **16.** $y = \ln(2x) - 8$

In Exercises 17 and 18, graph the function. Determine the asymptote of the function.

17. $y = \log_2(x + 1)$

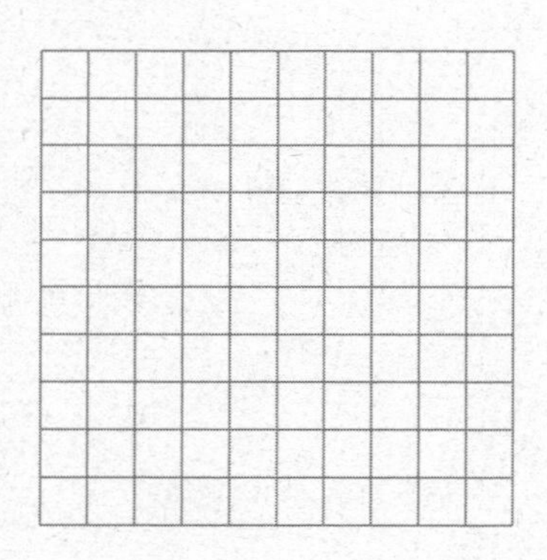

18. $y = \log_{1/2} x - 4$

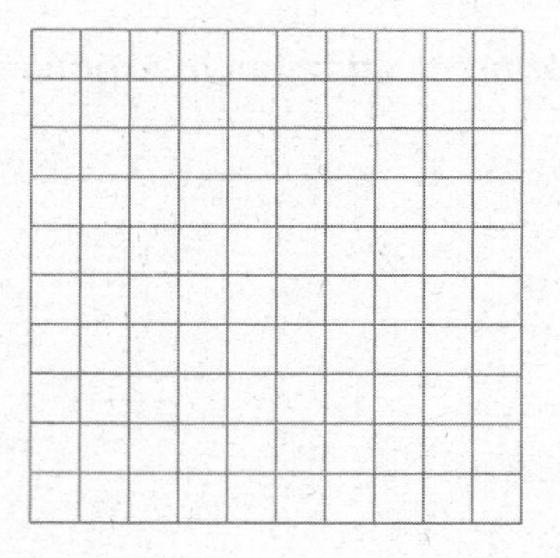

Name____ Date__________

6.4 Transformations of Exponential and Logarithmic Functions

For use with Exploration 6.4

Essential Question How can you transform the graphs of exponential and logarithmic functions?

1 EXPLORATION: Identifying Transformations

Work with a partner. Each graph shown is a transformation of the parent function

$$f(x) = e^x \qquad \text{or} \qquad f(x) = \ln x.$$

Match each function with its graph. Explain your reasoning. Then describe the transformation of f represented by g.

a. $g(x) = e^{x+2} - 3$ **b.** $g(x) = -e^{x+2} + 1$ **c.** $g(x) = e^{x-2} - 1$

d. $g(x) = \ln(x + 2)$ **e.** $g(x) = 2 + \ln x$ **f.** $g(x) = 2 + \ln(-x)$

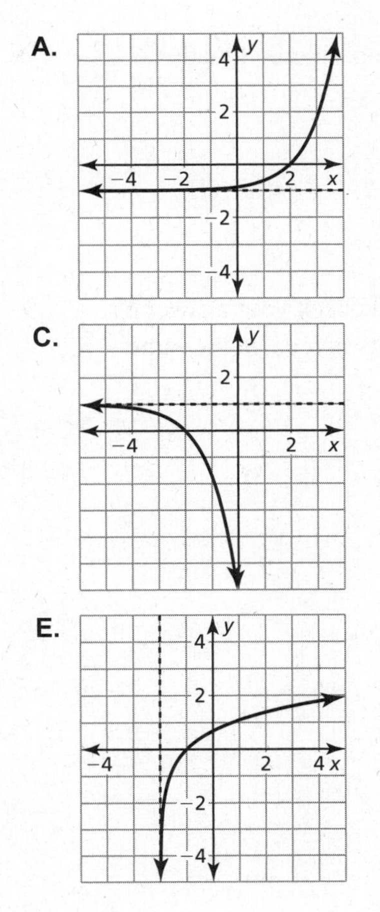

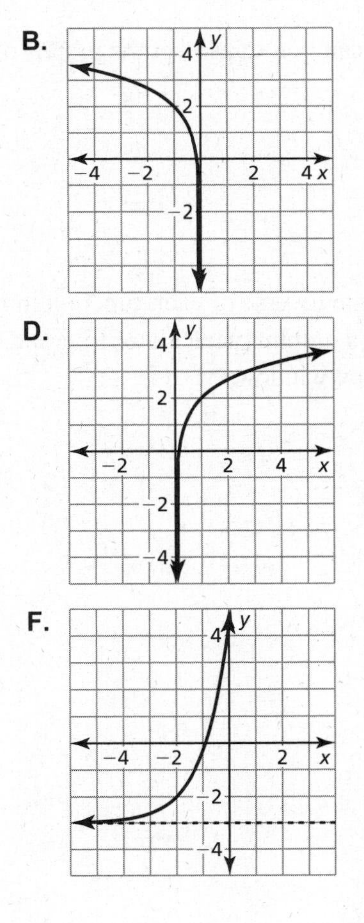

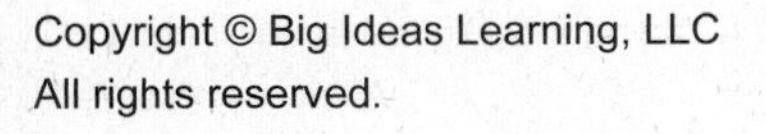

Name ____ Date__________

2 EXPLORATION: Characteristics of Graphs

Work with a partner. Determine the domain, range, and asymptote of each function in Exploration 1. Justify your answers.

Communicate Your Answer

3. How can you transform the graphs of exponential and logarithmic functions?

4. Find the inverse of each function in Exploration 1. Then check your answer by using a graphing calculator to graph each function and its inverse in the same viewing window.

Name____ Date__________

6.4 Notetaking with Vocabulary
For use after Lesson 6.4

In your own words, write the meaning of each vocabulary term.

exponential function

logarithmic function

transformations

Core Concepts

Transformation	$f(x)$ Notation	Examples	
Horizontal Translation Graph shifts left or right.	$f(x-h)$	$g(x) = 4^{x-3}$ $g(x) = 4^{x+2}$	3 units right 2 units left
Vertical Translation Graph shifts up or down.	$f(x) + k$	$g(x) = 4^x + 5$ $g(x) = 4^x - 1$	5 units up 1 unit down
Reflection Graph flips over x- or y-axis.	$f(-x)$ $-f(x)$	$g(x) = 4^{-x}$ $g(x) = -4^x$	over y-axis over x-axis
Horizontal Stretch or Shrink Graph stretches away from or shrinks toward y-axis	$f(ax)$	$g(x) = 4^{2x}$ $g(x) = 4^{x/2}$	shrink by $\frac{1}{2}$ stretch by 2
Vertical Stretch or Shrink Graph stretches away from or shrinks toward x-axis	$a \bullet f(x)$	$g(x) = 3(4^x)$ $g(x) = \frac{1}{4}(4^x)$	stretch by 3 shrink by $\frac{1}{4}$

Notes:

Name ______________________________ Date __________

6.4 Notetaking with Vocabulary (continued)

Transformation	$f(x)$ Notation	Examples	
Horizontal Translation Graph shifts left or right.	$f(x-h)$	$g(x) = \log(x-4)$ $g(x) = \log(x+7)$	4 units right 7 units left
Vertical Translation Graph shifts up or down.	$f(x)+k$	$g(x) = \log x + 3$ $g(x) = \log x - 1$	3 units up 1 unit down
Reflection Graph flips over x- or y-axis.	$f(-x)$ $-f(x)$	$g(x) = \log(-x)$ $g(x) = -\log x$	over y-axis over x-axis
Horizontal Stretch or Shrink Graph stretches away from or shrinks toward y-axis	$f(ax)$	$g(x) = \log(4x)$ $g(x) = \log\left(\frac{1}{3}x\right)$	shrink by $\frac{1}{4}$ stretch by 3
Vertical Stretch or Shrink Graph stretches away from or shrinks toward x-axis	$a \bullet f(x)$	$g(x) = 5 \log x$ $g(x) = \frac{2}{3} \log x$	stretch by 5 shrink by $\frac{2}{3}$

Notes:

Name____ Date__________

6.4 Notetaking with Vocabulary (continued)

Extra Practice

In Exercises 1–6, describe the transformation of *f* represented by *g*. Then graph each function.

1. $f(x) = 6^x, g(x) = 6^x + 6$

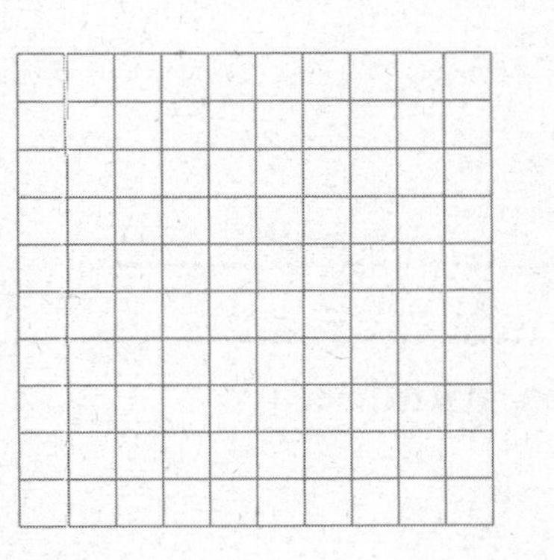

2. $f(x) = e^x, g(x) = e^{x-4}$

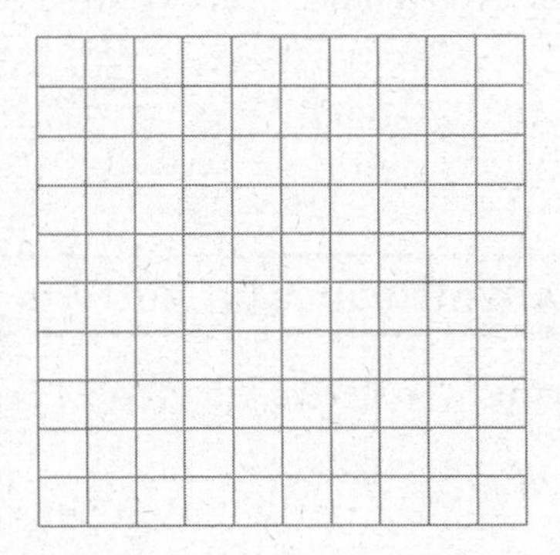

3. $f(x) = \log_5 x, g(x) = \frac{1}{2}\log_5(x + 7)$

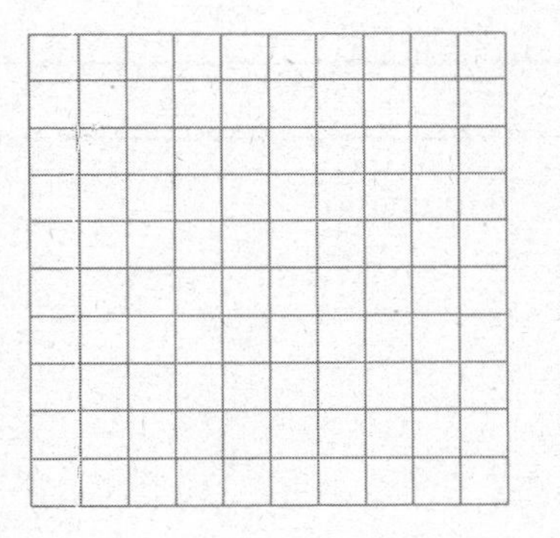

4. $f(x) = \log_{1/3} x, g(x) = \log_{1/3} x - \frac{4}{3}$

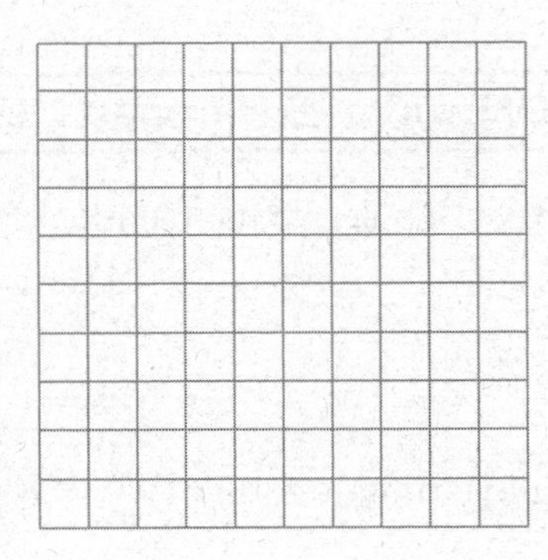

5. $f(x) = \left(\frac{1}{5}\right)^x, g(x) = \left(\frac{1}{5}\right)^{-3x} + 4$

6. $f(x) = \log x, g(x) = -3\log(x - 2)$

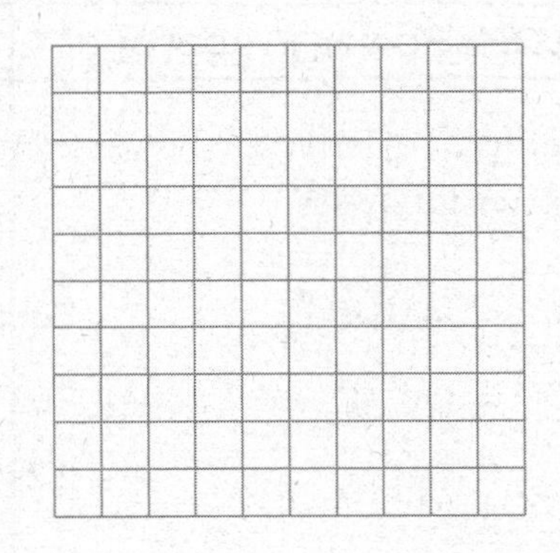

Name ______________________________ Date __________

6.5 Properties of Logarithms

For use with Exploration 6.5

Essential Question How can you use properties of exponents to derive properties of logarithms?

Let $x = \log_b m$ and $y = \log_b n.$

The corresponding exponential forms of these two equations are

$b^x = m$ and $b^y = n.$

EXPLORATION: Product Property of Logarithms

Work with a partner. To derive the Product Property, multiply m and n to obtain

$$mn = b^x b^y = b^{x+y}.$$

The corresponding logarithmic form of $mn = b^{x+y}$ is $\log_b mn = x + y$. So,

$\log_b mn =$ ________. Product Property of Logarithms

2 EXPLORATION: Quotient Property of Logarithms

Work with a partner. To derive the Quotient Property, divide m by n to obtain

$$\frac{m}{n} = \frac{b^x}{b^y} = b^{x-y}.$$

The corresponding logarithmic form of $\frac{m}{n} = b^{x-y}$ is $\log_b \frac{m}{n} = x - y$. So,

$\log_b \frac{m}{n} =$ ________. Quotient Property of Logarithms

3 EXPLORATION: Power Property of Logarithms

Work with a partner. To derive the Power Property, substitute b^x for m in the expression $\log_b m^n$, as follows.

$\log_b m^n = \log_b \left(b^x\right)^n$	Substitute b^x for m.
$= \log_b b^{nx}$	Power of a Power Property of Exponents
$= nx$	Inverse Property of Logarithms

Name____ Date__________

3 EXPLORATION: Power Property of Logarithms (continued)

So, substituting $\log_b m$ for x, you have

$\log_b m^n$ = ________.   Power Property of Logarithms

Communicate Your Answer

4. How can you use properties of exponents to derive properties of logarithms?

5. Use the properties of logarithms that you derived in Explorations 1–3 to evaluate each logarithmic expression.

a. $\log_4 16^3$

b. $\log_3 81^{-3}$

c. $\ln e^2 + \ln e^5$

d. $2 \ln e^6 - \ln e^5$

e. $\log_5 75 - \log_5 3$

f. $\log_4 2 + \log_4 32$

Name ______________________________ Date __________

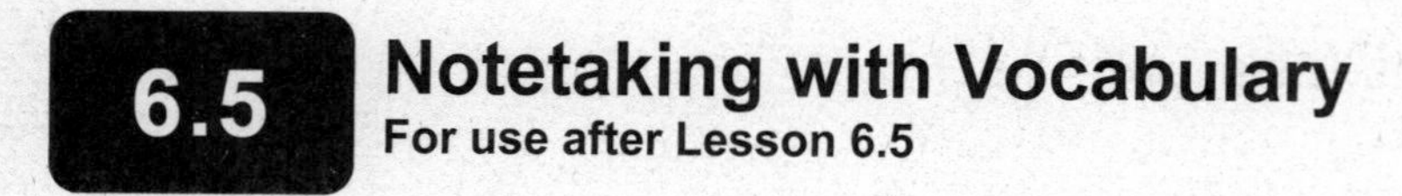

6.5 Notetaking with Vocabulary
For use after Lesson 6.5

In your own words, write the meaning of each vocabulary term.

base

properties of exponents

Core Concepts

Properties of Logarithms

Let b, m, and n be positive real numbers with $b \neq 1$.

Product Property $\log_b mn = \log_b m + \log_b n$

Quotient Property $\log_b \frac{m}{n} = \log_b m - \log_b n$

Power Property $\log_b m^n = n \log_b m$

Notes:

Change-of-Base Formula

If a, b, and c are positive real numbers with $b \neq 1$ and $c \neq 1$, then

$$\log_c a = \frac{\log_b a}{\log_b c}.$$

In particular, $\log_c a = \frac{\log a}{\log c}$ and $\log_c a = \frac{\ln a}{\ln c}$.

Notes:

Name____ Date__________

Extra Practice

In Exercises 1–4, use $\log_2 5 \approx 2.322$ and $\log_2 12 \approx 3.585$ to evaluate the logarithm.

1. $\log_2 60$
2. $\log_2 \frac{1}{144}$
3. $\log_2 \frac{12}{25}$
4. $\log_2 720$

In Exercises 5–8, expand the logarithmic expression.

5. $\log 10x$
6. $\ln 2x^6$
7. $\log_3 \frac{x^4}{3y^3}$
8. $\ln \sqrt[4]{3y^2}$

In Exercises 9–13, condense the logarithmic expression.

9. $\log_2 3 + \log_2 8$
10. $\log_5 4 - 2 \log_5 5$
11. $3 \ln 6x + \ln 4y$
12. $\log_2 625 - \log_2 125 + \frac{1}{3} \log_2 27$
13. $-\log_6 6 - \log_6 2y + 2 \log_6 3x$

In Exercises 14–17, use the change-of-base formula to evaluate the logarithm.

14. $\log_3 17$

15. $\log_9 294$

16. $\log_7 \frac{4}{9}$

17. $\log_6 \frac{1}{10}$

18. For a sound with intensity I (in watts per square meter), the loudness $L(I)$ of the sound (in decibels) is given by the function $L(I) = 10 \log \frac{I}{I_0}$, where I_0 is the intensity of a barely audible sound (about 10^{-12} watts per square meter). The intensity of the sound of a certain children's television show is half the intensity of the adult show that is on before it. By how many decibels does the loudness decrease?

19. Hick's Law states that given n equally probable choices, such as choices on a menu, the average human's reaction time T (in seconds) required to choose from those choices is approximately $T = a + b \bullet \log_2(n + 1)$ where a and b are constants. If $a = 4$ and $b = 1$, how much longer would it take a customer to choose what to eat from a menu of 40 items than from a menu of 10 items?

Name______________________________ Date__________

6.6 Solving Exponential and Logarithmic Equations

For use with Exploration 6.6

Essential Question How can you solve exponential and logarithmic equations?

1 EXPLORATION: Solving Exponential and Logarithmic Equations

Work with a partner. Match each equation with the graph of its related system of equations. Explain your reasoning. Then use the graph to solve the equation.

a. $e^x = 2$

b. $\ln x = -1$

c. $2^x = 3^{-x}$

d. $\log_4 x = 1$

e. $\log_5 x = \frac{1}{2}$

f. $4^x = 2$

A.

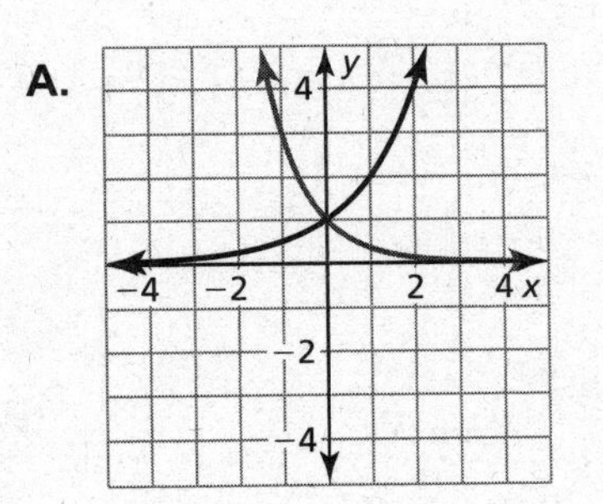

B.

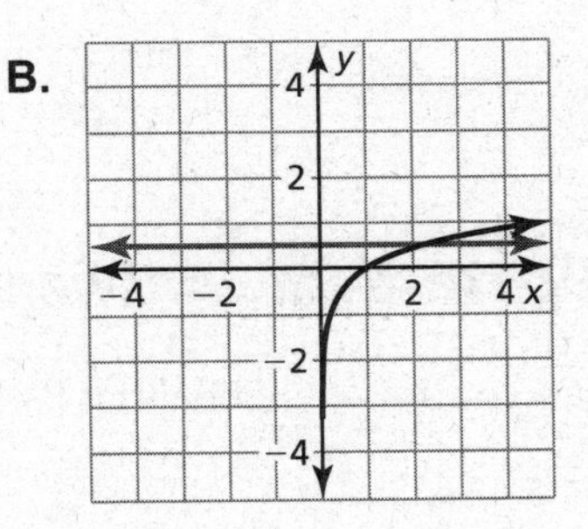

C.

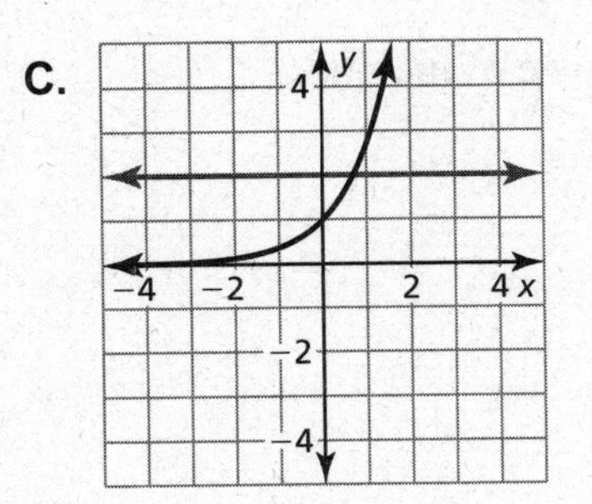

D.

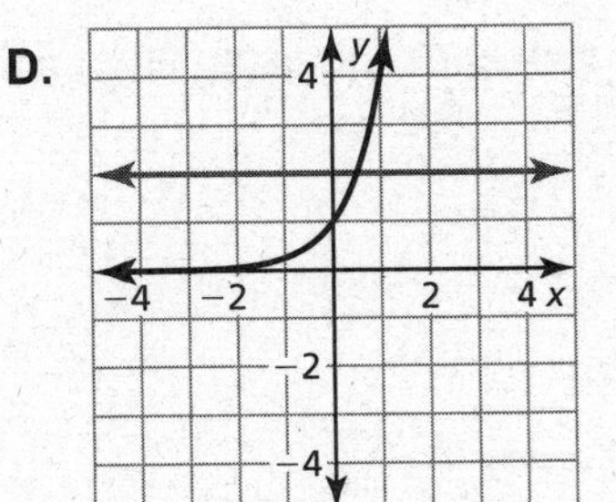

E.

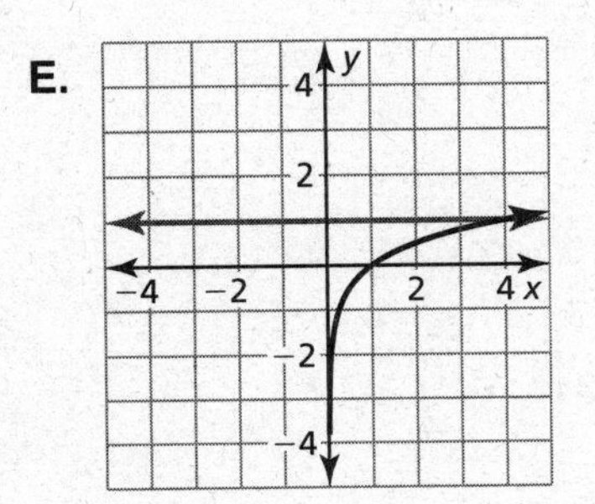

F.

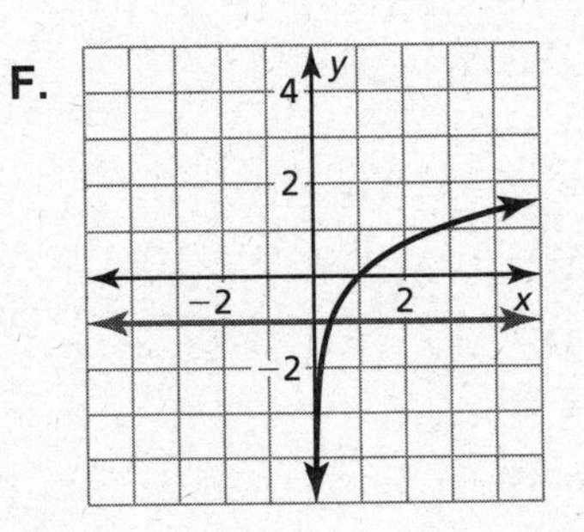

2 EXPLORATION: Solving Exponential and Logarithmic Equations

Go to *BigIdeasMath.com* for an interactive tool to investigate this exploration.

Work with a partner. Look back at the equations in Explorations 1(a) and 1(b). Suppose you want a more accurate way to solve the equations than using a graphical approach.

a. Show how you could use a *numerical approach* by creating a table. For instance, you might use a spreadsheet to solve the equations.

b. Show how you could use an *analytical approach*. For instance, you might try solving the equations by using the inverse properties of exponents and logarithms.

Communicate Your Answer

3. How can you solve exponential and logarithmic equations?

4. Solve each equation using any method. Explain your choice of method.

a. $16^x = 2$

b. $2^x = 4^{2x+1}$

c. $2^x = 3^{x+1}$

d. $\log x = \frac{1}{2}$

e. $\ln x = 2$

f. $\log_3 x = \frac{3}{2}$

Name____ Date__________

6.6 Notetaking with Vocabulary

For use after Lesson 6.6

In your own words, write the meaning of each vocabulary term.

exponential equations

logarithmic equations

Core Concepts

Property of Equality for Exponential Equations

Algebra If b is a positive real number other than 1, then $b^x = b^y$ if and only if $x = y$.

Example If $3^x = 3^5$, then $x = 5$. If $x = 5$, then $3^x = 3^5$.

Notes:

Property of Equality for Logarithmic Equations

Algebra If b, x, and y are positive real numbers with $b \neq 1$, then $\log_b x = \log_b y$ if and only if $x = y$.

Example If $\log_2 x = \log_2 7$, then $x = 7$. If $x = 7$, then $\log_2 x = \log_2 7$.

Notes:

Name ______________________________ Date ________

Extra Practice

In Exercises 1–6, solve the equation.

1. $5^{2x+4} = 5^{5x-8}$

2. $4^{2x-1} = 8^{x+2}$

3. $3^{x+3} = 5$

4. $\left(\frac{1}{5}\right)^{3x-2} = \sqrt{25^x}$

5. $12e^{1-x} = 500$

6. $-14 + 3e^x = 11$

In Exercises 7–11, solve the equation. Check for extraneous solutions.

7. $2 = \log_3(4x)$

8. $\ln(x^2 + 3) = \ln(4)$

9. $\log_8(x^2 - 5) = \frac{2}{3}$

10. $\ln x + \ln(x + 2) = \ln(x + 6)$

11. $\log_2(x + 5) - \log_2(x - 2) = 3$

12. Solve the inequality $\log x \le \frac{1}{2}$.

13. Your parents buy juice for your graduation party and leave it in their hot car. When they take the cans out of the car and move them to the basement, the temperature of the juice is 80°F. The room temperature of the basement is 60°F, and the cooling rate of the juice is $r = 0.0147$. Using Newton's Law of Cooling, how long will it take to cool the juice to 63°F?

14. Earthquake intensity is measured by the formula $R = \log\left(\frac{I}{I_0}\right)$ where R is the Richter scale rating of an earthquake, I is the intensity of the earthquake, and I_0 is the intensity of the smallest detectable wave. In 1906, an earthquake in San Francisco had an estimated measure of 7.8 on the Richter scale. In the same year, another earthquake had an intensity level four times stronger than the San Francisco earthquake giving it a Richter scale rating of $R_2 = \log\left(\frac{4I}{I_0}\right)$. What was the Richter scale rating on a scale of 1–10 of the other earthquake?

Name ______________________________ Date __________

6.7 Modeling with Exponential and Logarithmic Functions

For use with Exploration 6.7

Essential Question How can you recognize polynomial, exponential, and logarithmic models?

1 EXPLORATION: Recognizing Different Types of Models

Go to *BigIdeasMath.com* for an interactive tool to investigate this exploration.

Work with a partner. Match each type of model with the appropriate scatter plot. Use a regression program to find a model that fits the scatter plot.

a. linear (positive slope) **b.** linear (negative slope) **c.** quadratic

d. cubic **e.** exponential **f.** logarithmic

A.

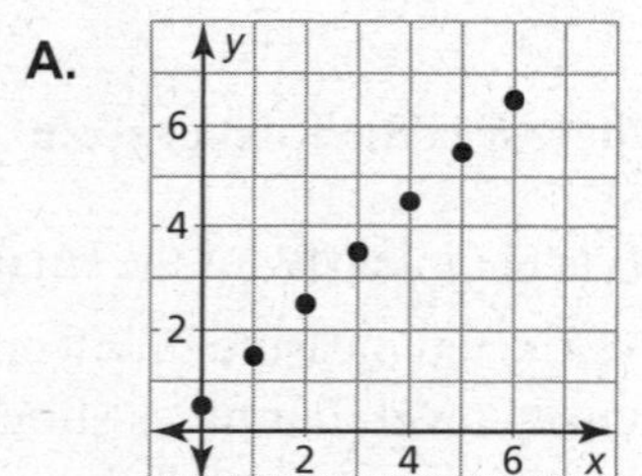

B.

C.

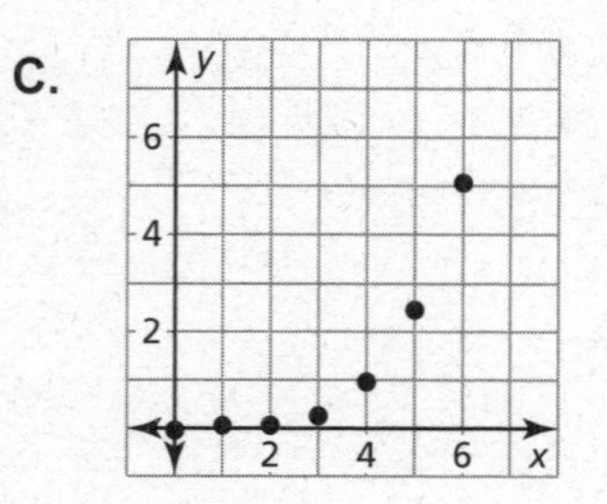

D.

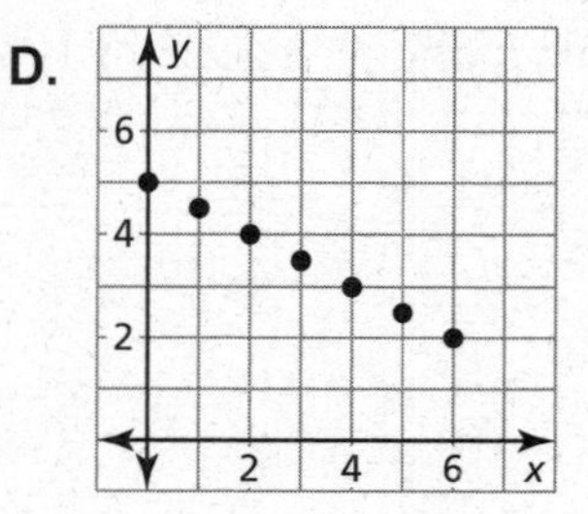

E.

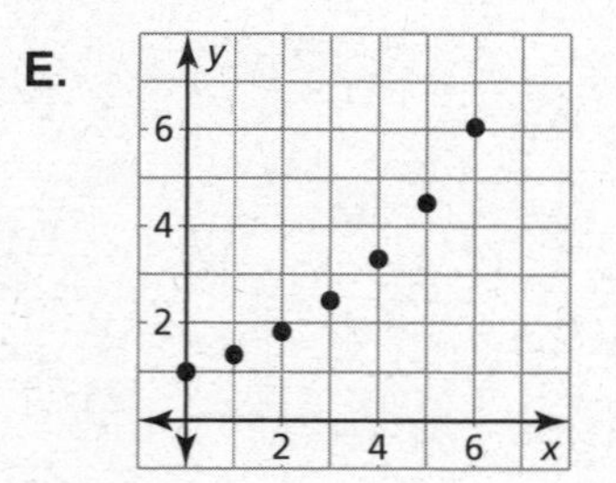

F.

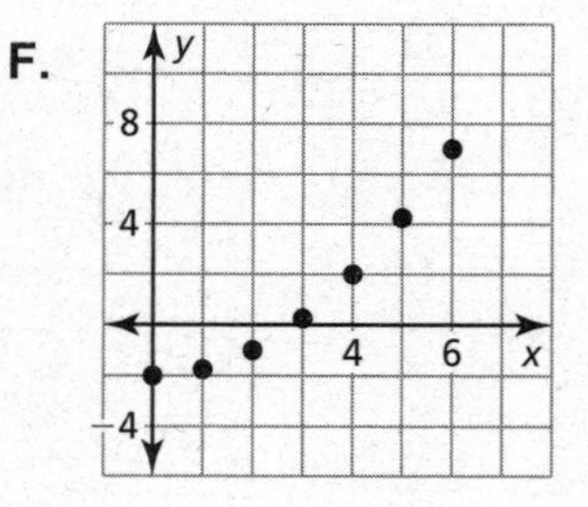

2 EXPLORATION: Exploring Gaussian and Logistic Models

Go to *BigIdeasMath.com* for an interactive tool to investigate this exploration.

Work with a partner. Two common types of functions that are related to exponential functions are given. Use a graphing calculator to graph each function. Then determine the domain, range, intercept, and asymptote(s) of the function.

a. Gaussian Function: $f(x) = e^{-x^2}$

b. Logistic Function: $f(x) = \dfrac{1}{1 + e^{-x}}$

Communicate Your Answer

3. How can you recognize polynomial, exponential, and logarithmic models?

4. Use the Internet or some other reference to find real-life data that can be modeled using one of the types given in Exploration 1. Create a table and a scatter plot of the data. Then use a regression program to find a model that fits the data.

Name ______________________________ Date __________

6.7 Notetaking with Vocabulary
For use after Lesson 6.7

In your own words, write the meaning of each vocabulary term.

finite differences

common ratio

point-slope form

Notes:

Name______________________________ Date__________

6.7 Notetaking with Vocabulary (continued)

Extra Practice

In Exercises 1 and 2, determine the type of function represented by the table. Explain your reasoning.

1.

x	6	7	8	9	10	11
y	34	47	62	79	98	119

2.

x	-5	-3	-1	1	3	5
y	$\frac{1}{5}$	$\frac{3}{5}$	$\frac{9}{5}$	$\frac{27}{5}$	$\frac{81}{5}$	$\frac{243}{5}$

In Exercises 3–6, write an exponential function $y = ab^x$ whose graph passes through the given points.

3. $(1, 12), (3, 108)$

4. $(-1, 2), (3, 32)$

5. $(2, 9), (4, 324)$

6. $(-2, 2), (1, 0.25)$

Name ______________________________ Date __________

6.7 Notetaking with Vocabulary (continued)

7. An Olympic swimmer starts selling a new type of goggles. The table shows the number y of goggles sold during a 6-month period.

Months, *x*	1	2	3	4	5	6
Goggles sold, *y*	28	47	64	79	97	107

a. Create a scatterplot of the data.

b. Create a scatterplot of the data pairs $(x, \ln y)$ to show that an exponential model should be a good fit for the original data pairs (x, y). Write a function that models the data.

c. Use a graphing calculator to write an exponential model for the data.

d. Use each model to predict the number of goggles sold after 1 year.

Name____ Date__________

Chapter 7 Maintaining Mathematical Proficiency

Evaluate.

1. $\frac{2}{3} + \frac{2}{3}$

2. $\frac{1}{5} + \frac{1}{4}$

3. $-\frac{5}{6} + \frac{3}{4}$

4. $\frac{9}{11} - \frac{2}{11}$

5. $\frac{1}{5} - \frac{7}{10}$

6. $\frac{5}{8} - \frac{1}{6}$

7. $-\frac{3}{8} + \frac{2}{9} - \frac{1}{2}$

8. $\frac{3}{4} - \left(-\frac{1}{8}\right)$

9. $\frac{13}{18} + \frac{2}{9} - \frac{1}{2}$

Simplify.

10. $\frac{\frac{2}{3}}{\frac{8}{15}}$

11. $\frac{\frac{1}{6}}{-\frac{2}{3}}$

12. $\frac{\frac{3}{4}}{12}$

13. $\frac{1}{\frac{1}{5} + \frac{2}{5}}$

14. $\frac{2}{\frac{4}{9} - \frac{2}{3}}$

15. $\frac{\frac{1}{2} + \frac{1}{5}}{\frac{7}{10} - \frac{2}{5}}$

Name ______________________________ Date __________

7.1 Inverse Variation

For use with Exploration 7.1

Essential Question How can you recognize when two quantities vary directly or inversely?

1 EXPLORATION: Recognizing Direct Variation

Go to *BigIdeasMath.com* for an interactive tool to investigate this exploration.

Work with a partner. You hang different weights from the same spring.

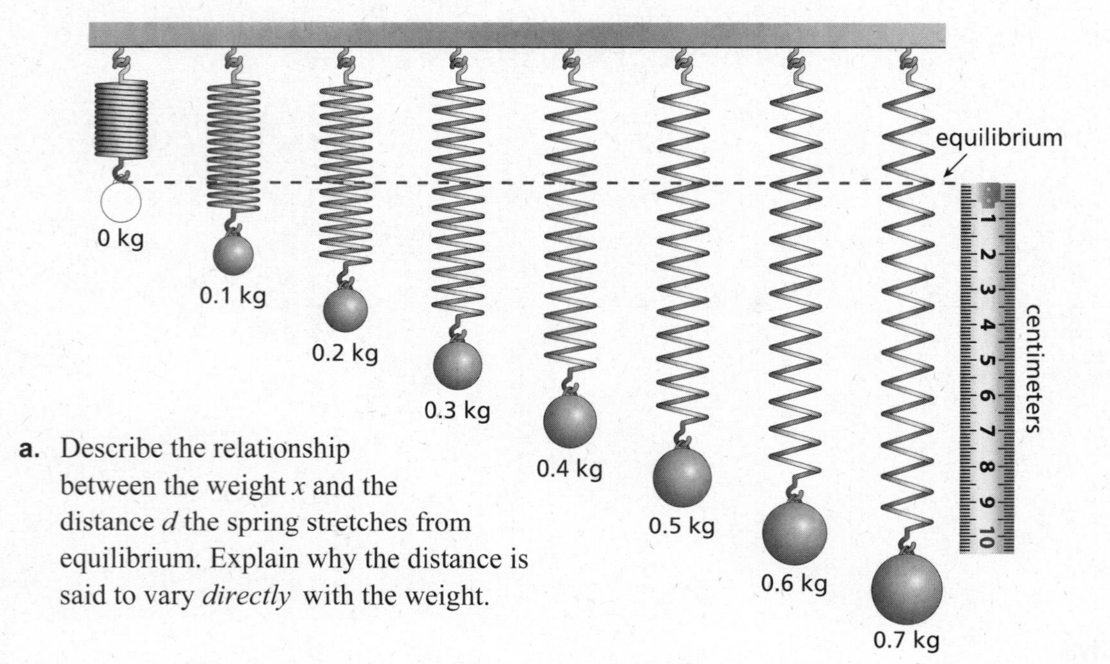

a. Describe the relationship between the weight x and the distance d the spring stretches from equilibrium. Explain why the distance is said to vary *directly* with the weight.

b. Estimate the values of d from the figure. Then draw a scatter plot of the data. What are the characteristics of the graph?

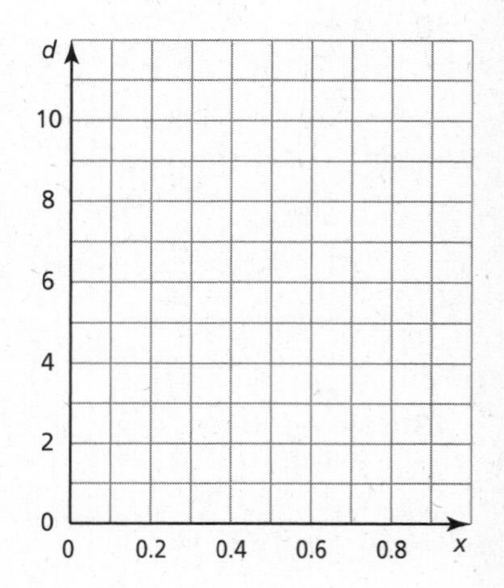

c. Write an equation that represents d as a function of x.

d. In physics, the relationship between d and x is described by *Hooke's Law*. How would you describe Hooke's Law?

2 EXPLORATION: Recognizing Inverse Variation

Go to *BigIdeasMath.com* for an interactive tool to investigate this exploration.

Work with a partner. The table shows the length x (in inches) and the width y (in inches) of a rectangle. The area of each rectangle is 64 square inches.

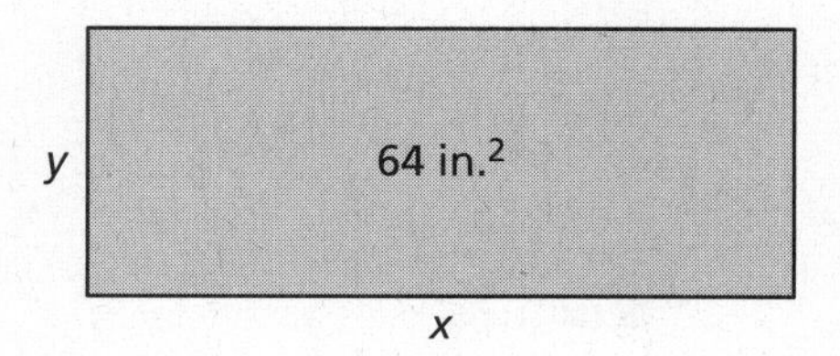

x	y
1	
2	
4	
8	
16	
32	
64	

a. Complete the table.

b. Describe the relationship between x and y. Explain why y is said to vary *inversely* with x.

c. Draw a scatter plot of the data. What are the characteristics of the graph?

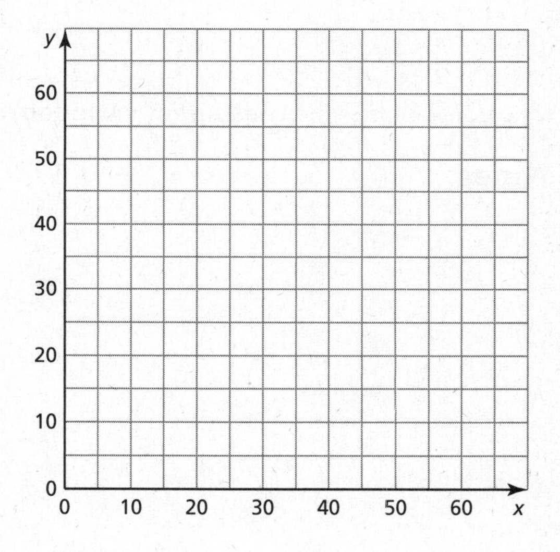

d. Write an equation that represents y as a function of x.

Communicate Your Answer

3. How can you recognize when two quantities vary directly or inversely?

4. Does the flapping rate of the wings of a bird vary directly or inversely with the length of its wings? Explain your reasoning.

Name ____ Date __________

7.1 Notetaking with Vocabulary

For use after Lesson 7.1

In your own words, write the meaning of each vocabulary term.

inverse variation

constant of variation

Core Concepts

Inverse Variation

Two variables x and y show **inverse variation** when they are related as follows:

$$y = \frac{a}{x}, \quad a \neq 0$$

The constant a is the **constant of variation,** and y is said to *vary inversely* with x.

Notes:

Extra Practice

In Exercises 1–9, tell whether *x* and *y* show *direct variation*, *inverse variation*, or *neither*.

1. $3xy = 1$

2. $\frac{5}{x} = y$

3. $x + 11 = y$

4. $x + y = -2$

5. $\frac{4}{5}x = y$

6. $x - 8y = 1$

7. $\frac{x}{7} = y$

8. $6xy = 0$

9. $\frac{y}{9x} = 1$

In Exercises 10–12, tell whether *x* and *y* show *direct variation*, *inverse variation*, or *neither*.

10.

x	2	4	6	8	10
y	4	16	36	64	100

11.

x	1	5	8	20	50
y	5	1	0.625	0.25	0.1

12.

x	2	5	8.4	12	15
y	0.5	1.25	2.1	3	3.75

7.1 Notetaking with Vocabulary (continued)

In Exercises 13–16, the variables *x* and *y* vary inversely. Use the given values to write an equation relating *x* and *y*. Then find *y* when $x = 5$.

13. $x = 2,\ y = 2$

14. $x = 6,\ y = 3$

15. $x = 20,\ y = \frac{7}{20}$

16. $x = \frac{10}{9},\ y = \frac{3}{2}$

17. When temperature is held constant, the volume V of a gas is inversely proportional to the pressure P of the gas on its container. A pressure of 32 pounds per square inch results in a volume of 20 cubic feet. What is the pressure if the volume becomes 10 cubic feet?

18. The time t (in days) that it takes to harvest a field varies inversely with the number n of farm workers. A farmer can harvest his crop in 20 days with 7 farm workers. How long will it take to harvest the crop if he hires 10 farm workers?

Name____ Date__________

7.2 Graphing Rational Functions

For use with Exploration 7.2

Essential Question

What are some of the characteristics of the graph of a rational function?

The parent function for rational functions with a linear numerator and a linear denominator is

$$f(x) = \frac{1}{x}. \qquad \text{Parent function}$$

The graph of this function, shown at the right, is a *hyperbola*.

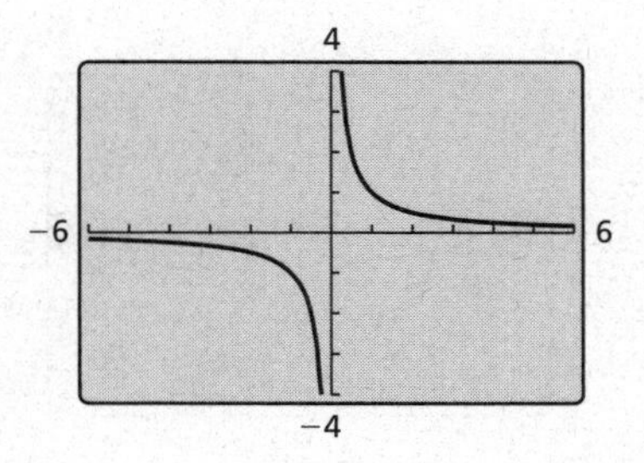

1 EXPLORATION: Identifying Graphs of Rational Functions

Work with a partner. Each function is a transformation of the graph of the parent function $f(x) = \frac{1}{x}$. Match the function with its graph. Explain your reasoning. Then describe the transformation.

a. $g(x) = \frac{1}{x - 1}$ **b.** $g(x) = \frac{-1}{x - 1}$ **c.** $g(x) = \frac{x + 1}{x - 1}$

d. $g(x) = \frac{x - 2}{x + 1}$ **e.** $g(x) = \frac{x}{x + 2}$ **f.** $g(x) = \frac{-x}{x + 2}$

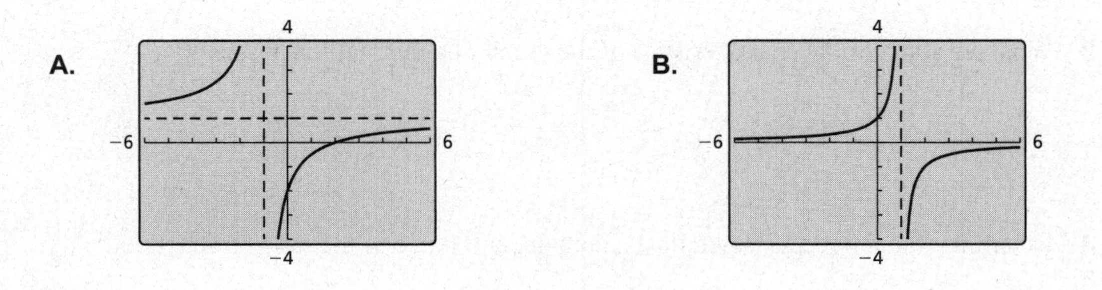

Name ______________________ Date ________

1 EXPLORATION: Identifying Graphs of Rational Functions (continued)

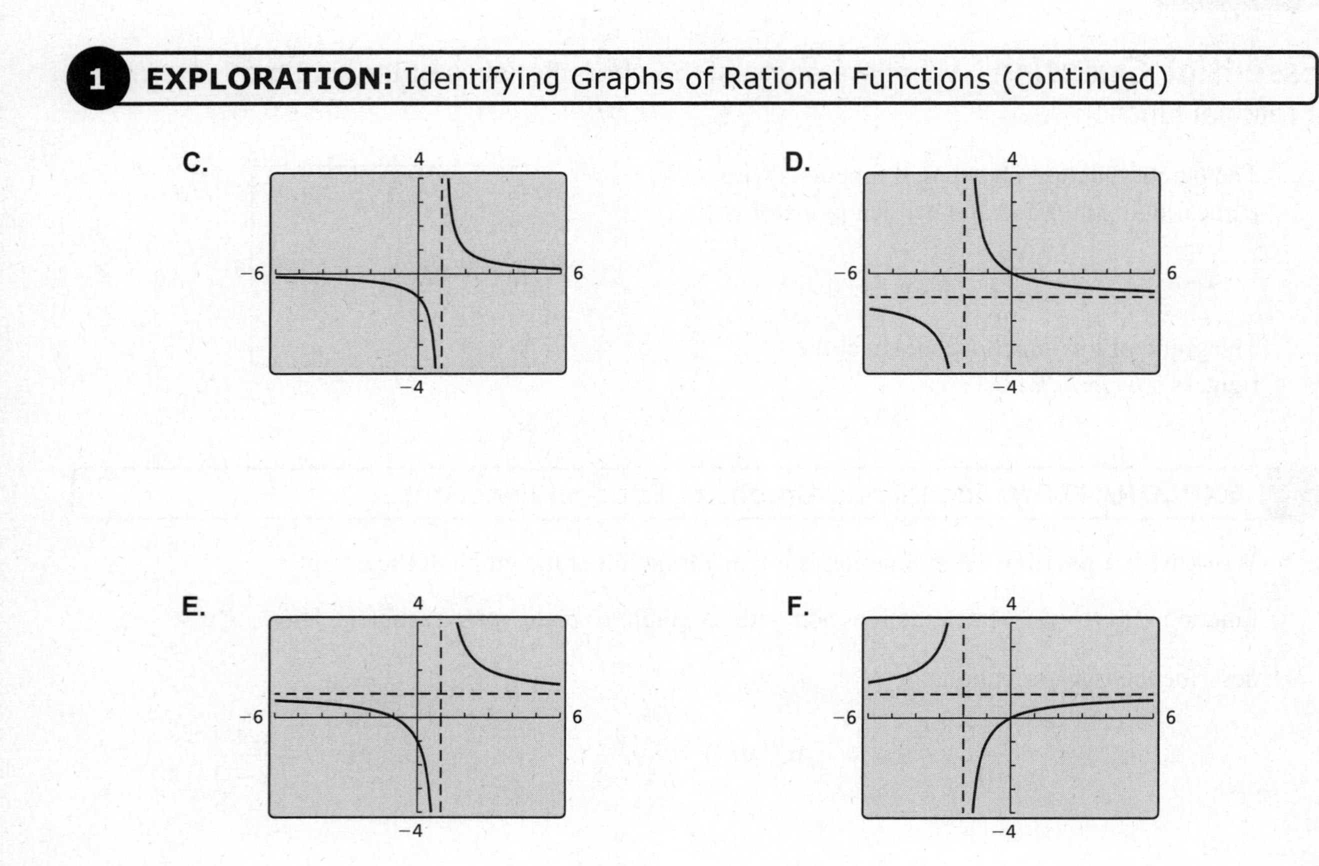

Communicate Your Answer

2. What are some of the characteristics of the graph of a rational function?

3. Determine the intercepts, asymptotes, domain, and range of the rational function

$$g(x) = \frac{x - a}{x - b}.$$

Name____ Date__________

7.2 Notetaking with Vocabulary

For use after Lesson 7.2

In your own words, write the meaning of each vocabulary term.

rational function

Core Concepts

Parent Function for Simple Rational Functions

The graph of the parent function $f(x) = \frac{1}{x}$ is a *hyperbola*, which consists of two symmetrical parts called branches. The domain and range are all nonzero real numbers.

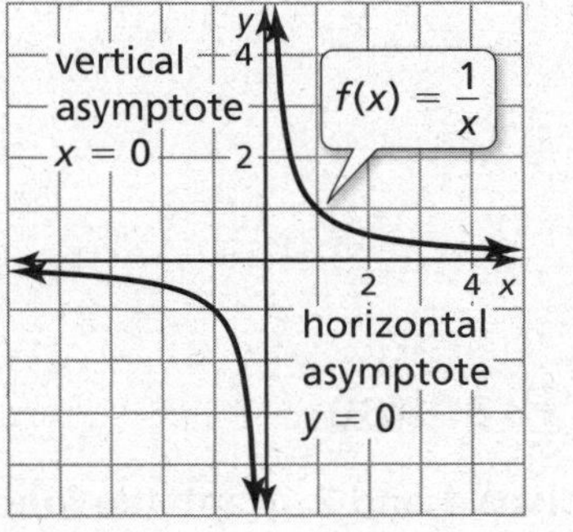

Any function of the form $g(x) = \frac{a}{x} (a \neq 0)$ has the same asymptotes, domain, and range as the function $f(x) = \frac{1}{x}$.

Notes:

Name ______________________________ Date __________

7.2 Notetaking with Vocabulary (continued)

Graphing Translations of Simple Rational Functions

To graph a rational function of the form $y = \frac{a}{x - h} + k$, follow these steps:

Step 1 Draw the asymptotes $x = h$ and $y = k$.

Step 2 Plot points to the left and to the right of the vertical asymptote.

Step 3 Draw the two branches of the hyperbola so that they pass through the plotted points and approach the asymptotes.

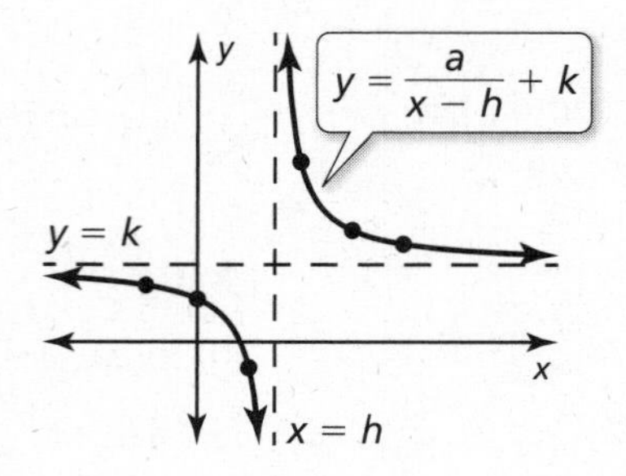

Notes:

Extra Practice

In Exercises 1 and 2, graph the function. Compare the graph with the graph of $f(x) = \frac{1}{x}$.

1. $g(x) = \frac{0.25}{x}$

2. $h(x) = \frac{-2}{x}$

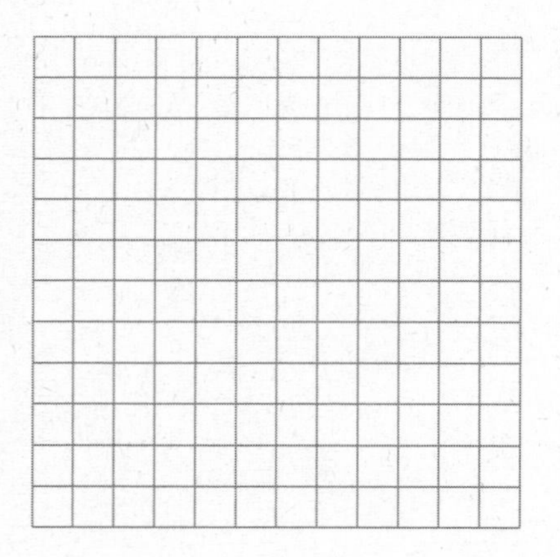

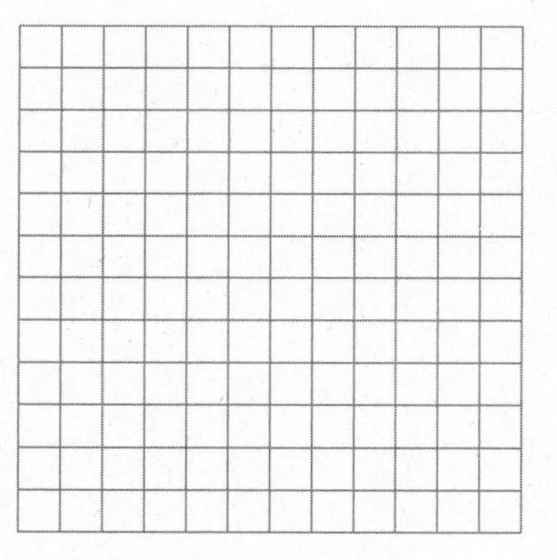

Name____ Date__________

7.2 Notetaking with Vocabulary (continued)

In Exercises 3 and 4, graph the function. State the domain and range.

3. $k(x) = \dfrac{1}{x - 3} + 5$

4. $m(x) = \dfrac{-3}{x} - 4$

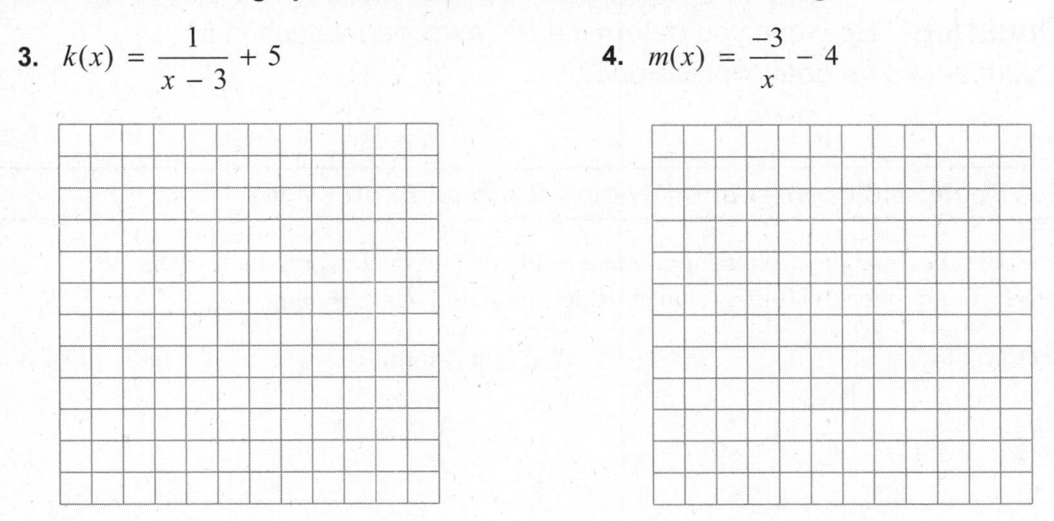

In Exercises 5 and 6, rewrite the function in the form $g(x) = \dfrac{a}{x - h} + k$. Graph the function. Describe the graph of g as a transformation of the graph of $f(x) = \dfrac{a}{x}$.

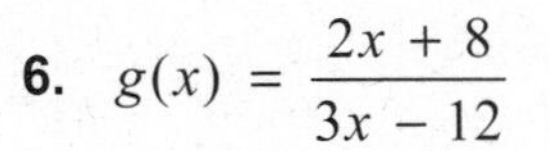

5. $g(x) = \dfrac{x + 2}{x - 5}$

6. $g(x) = \dfrac{2x + 8}{3x - 12}$

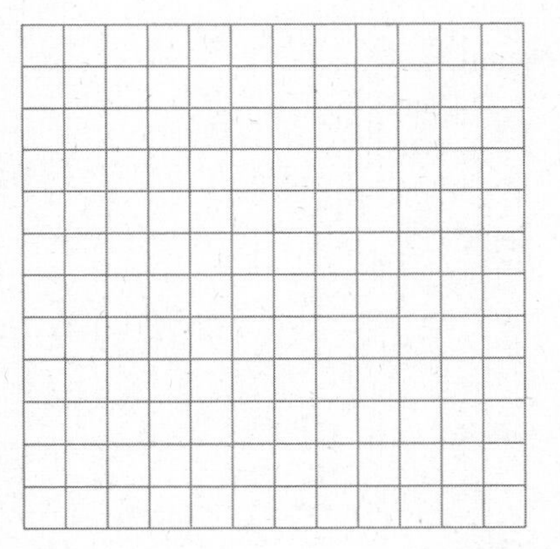

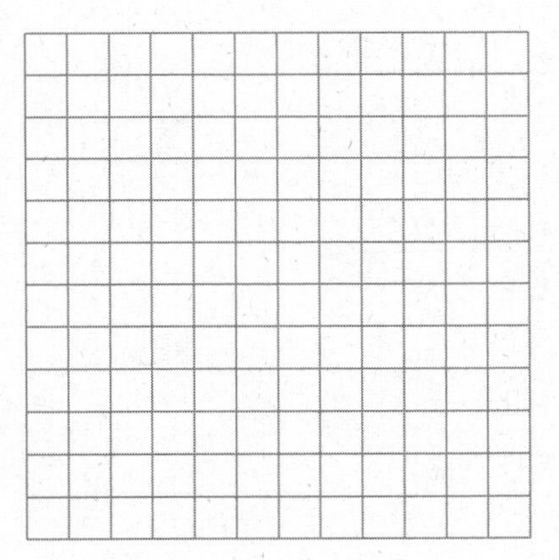

Name ____ Date __________

7.3 Multiplying and Dividing Rational Expressions

For use with Exploration 7.3

Essential Question How can you determine the excluded values in a product or quotient of two rational expressions?

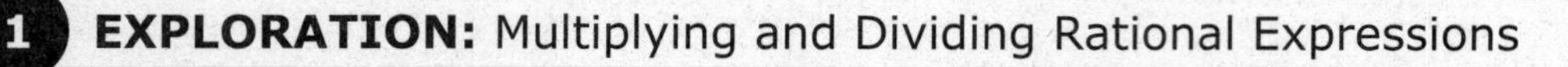

1 EXPLORATION: Multiplying and Dividing Rational Expressions

Work with a partner. Find the product or quotient of the two rational expressions. Then match the product or quotient with its excluded values. Explain your reasoning.

Product or Quotient	Excluded Values
a. $\frac{1}{x-1} \bullet \frac{x-2}{x+1} =$	**A.** $-1, 0,$ and 2
b. $\frac{1}{x-1} \bullet \frac{-1}{x-1} =$	**B.** -2 and 1
c. $\frac{1}{x-2} \bullet \frac{x-2}{x+1} =$	**C.** $-2, 0,$ and 1
d. $\frac{x+2}{x-1} \bullet \frac{-x}{x+2} =$	**D.** -1 and 2
e. $\frac{x}{x+2} \div \frac{x+1}{x+2} =$	**E.** $-1, 0,$ and 1
f. $\frac{x}{x-2} \div \frac{x+1}{x} =$	**F.** -1 and 1
g. $\frac{x}{x+2} \div \frac{x}{x-1} =$	**G.** -2 and -1
h. $\frac{x+2}{x} \div \frac{x+1}{x-1} =$	**H.** 1

Name____ Date__________

2 EXPLORATION: Writing a Product or Quotient

Work with a partner. Write a product or quotient of rational expressions that has the given excluded values. Justify your answer.

a. -1

b. -1 and 3

c. -1, 0, and 3

Communicate Your Answer

3. How can you determine the excluded values in a product or quotient of two rational expressions?

4. Is it possible for the product or quotient of two rational expressions to have *no* excluded values? Explain your reasoning. If it is possible, give an example.

Name ______________________________ Date __________

7.3 Notetaking with Vocabulary

For use after Lesson 7.3

In your own words, write the meaning of each vocabulary term.

rational expression

simplified form of a rational expression

Core Concepts

Simplifying Rational Expressions

Let a, b, and c be expressions with $b \neq 0$ and $c \neq 0$.

Property $\dfrac{a\cancel{c}}{b\cancel{c}} = \dfrac{a}{b}$ Divide out common factor c.

Examples $\dfrac{15}{65} = \dfrac{3 \bullet \cancel{5}}{13 \bullet \cancel{5}} = \dfrac{3}{13}$ Divide out common factor 5.

$\dfrac{4\cancel{(x + 3)}}{(x + 3)\cancel{(x + 3)}} = \dfrac{4}{x + 3}$ Divide out common factor $x + 3$.

Notes:

Name______________________________ Date__________

7.3 Notetaking with Vocabulary (continued)

Multiplying Rational Expressions

Let a, b, c, and d be expressions with $b \neq 0$ and $d \neq 0$.

Property $\frac{a}{b} \bullet \frac{c}{d} = \frac{ac}{bd}$ Simplify $\frac{ac}{bd}$ if possible.

____

Example

$$\frac{5x^2}{2xy^2} \bullet \frac{6xy^3}{10y} = \frac{30x^3y^3}{20xy^3} = \frac{\cancel{10} \bullet 3 \bullet \cancel{x} \bullet x^2 \bullet \cancel{y^3}}{\cancel{10} \bullet 2 \bullet \cancel{x} \bullet \cancel{y^3}} = \frac{3x^2}{2}, x \neq 0, y \neq 0$$

Notes:

Dividing Rational Expressions

Let a, b, c, and d be expressions with $b \neq 0$, $c \neq 0$, and $d \neq 0$.

Property $\frac{a}{b} \div \frac{c}{d} = \frac{a}{b} \bullet \frac{d}{c} = \frac{ad}{bc}$ Simplify $\frac{ad}{bc}$ if possible.

____

Example $\frac{7}{x+1} \div \frac{x+2}{2x-3} = \frac{7}{x+1} \bullet \frac{2x-3}{x+2} = \frac{7(2x-3)}{(x+1)(x+2)}, x \neq \frac{3}{2}$

Notes:

Name ______________________________ Date __________

7.3 Notetaking with Vocabulary (continued)

Extra Practice

In Exercises 1–4, simplify the expression, if possible.

1. $\dfrac{2x^3 - 8x^2}{6x^2}$

2. $\dfrac{5xy^3 - 2x^2y^2}{x^2y^2}$

3. $\dfrac{x^2 - 5x + 4}{x^2 - 2x + 1}$

4. $\dfrac{x^3 + 3x^2}{x^2 - 5x - 24}$

In Exercises 5–10, find the product or the quotient.

5. $\dfrac{3xy}{xy^2} \bullet \dfrac{y}{2x}$

6. $\dfrac{x + y}{7xy} \div \dfrac{4x}{y}$

7. $\dfrac{x(x + 1)}{x - 2} \div \dfrac{(x + 1)(x - 6)}{(x - 6)(x - 9)}$

8. $\dfrac{x^2 - 2x - 3}{x^2 - 1} \bullet \dfrac{x^2 - 2x - 63}{x^2 + 4x - 21}$

9. $\dfrac{x^2 - 2x}{x + 7} \bullet \dfrac{x^3 + 8}{x^3 - 4x}$

10. $\dfrac{x^2 + 2x - 15}{x^2 - 3x - 40} \div \dfrac{x^2 + 8x - 9}{x^2 + x - 72}$

Name____ Date__________

7.4 Adding and Subtracting Rational Expressions

For use with Exploration 7.4

Essential Question How can you determine the domain of the sum or difference of two rational expressions?

1 EXPLORATION: Adding and Subtracting Rational Expressions

Work with a partner. Find the sum or difference of the two rational expressions. Then match the sum or difference with its domain. Explain your reasoning.

Sum or Difference	Domain
a. $\frac{1}{x-1} + \frac{3}{x-1} =$	**A.** all real numbers except -2
b. $\frac{1}{x-1} + \frac{1}{x} =$	**B.** all real numbers except -1 and 1
c. $\frac{1}{x-2} + \frac{1}{2-x} =$	**C.** all real numbers except 1
d. $\frac{1}{x-1} + \frac{-1}{x+1} =$	**D.** all real numbers except 0
e. $\frac{x}{x+2} - \frac{x+1}{2+x} =$	**E.** all real numbers except -2 and 1
f. $\frac{x}{x-2} - \frac{x+1}{x} =$	**F.** all real numbers except 0 and 1
g. $\frac{x}{x+2} - \frac{x}{x-1} =$	**G.** all real numbers except 2
h. $\frac{x+2}{x} - \frac{x+1}{x} =$	**H.** all real numbers except 0 and 2

7.4 Adding and Subtracting Rational Expressions (continued)

2 EXPLORATION: Writing a Sum or Difference

Work with a partner. Write a sum or difference of rational expressions that has the given domain. Justify your answer.

a. all real numbers except -1

b. all real numbers except -1 and 3

c. all real numbers except -1, 0, and 3

Communicate Your Answer

3. How can you determine the domain of the sum or difference of two rational expressions?

4. Your friend found a sum as follows. Describe and correct the error(s).

$$\frac{x}{x+4} + \frac{3}{x-4} = \frac{x+3}{2x}$$

Name____ Date__________

7.4 Notetaking with Vocabulary

For use after Lesson 7.4

In your own words, write the meaning of each vocabulary term.

complex fraction

Core Concepts

Adding or Subtracting with Like Denominators

Let a, b, and c be expressions with $c \neq 0$.

Addition

$$\frac{a}{c} + \frac{b}{c} = \frac{a + b}{c}$$

Subtraction

$$\frac{a}{c} - \frac{b}{c} = \frac{a - b}{c}$$

Notes:

Adding or Subtracting with Unlike Denominators

Let a, b, c, and d be expressions with $c \neq 0$ and $d \neq 0$.

Addition

$$\frac{a}{c} + \frac{b}{d} = \frac{ad}{cd} + \frac{bc}{cd} = \frac{ad + bc}{cd}$$

Subtraction

$$\frac{a}{c} - \frac{b}{d} = \frac{ad}{cd} - \frac{bc}{cd} = \frac{ad - bc}{cd}$$

Notes:

Name ______________________________ Date __________

7.4 Notetaking with Vocabulary (continued)

Simplifying Complex Fractions

Method 1 If necessary, simplify the numerator and denominator by writing each as a single fraction. Then divide by multiplying the numerator by the reciprocal of the denominator.

Method 2 Multiply the numerator and the denominator by the LCD of *every* fraction in the numerator and denominator. Then simplify.

Notes:

Extra Practice

In Exercises 1–4, find the sum or difference.

1. $\frac{1}{x-1} - \frac{5}{x-1}$

2. $\frac{4x}{3x-5} + \frac{x}{3x-5}$

3. $\frac{6x}{x+4} + \frac{24}{x+4}$

4. $\frac{2x^2}{x-7} - \frac{14x}{x-7}$

Name____ Date__________

In Exercises 5–7, find the least common multiple of the expressions.

5. $9x^3, 3x^2 - 21x$

6. $x + 5, 2x^2 + 11x + 5$

7. $x^2 + 5x + 6, x^2 - 3x - 18$

In Exercises 8–11, find the sum or the difference.

8. $\dfrac{3}{2x} + \dfrac{11}{5x}$

9. $\dfrac{15}{x - 2} + \dfrac{3}{x + 8}$

10. $\dfrac{3x}{2x + 1} + \dfrac{10}{2x^2 - 5x - 3}$

11. $\dfrac{x}{x - 7} - \dfrac{2}{x + 1} - \dfrac{8x}{x^2 - 6x - 7}$

In Exercises 12 and 13, simplify the complex fraction.

12. $\dfrac{\dfrac{x}{10} - 3}{5 + \dfrac{1}{x}}$

13. $\dfrac{\dfrac{12}{x^2 - 7x - 44}}{\dfrac{2}{x - 11} + \dfrac{1}{x + 4}}$

Name ______________________________ Date __________

7.5 Solving Rational Equations

For use with Exploration 7.5

Essential Question How can you solve a rational equation?

1 EXPLORATION: Solving Rational Equations

Work with a partner. Match each equation with the graph of its related system of equations. Explain your reasoning. Then use the graph to solve the equation.

a. $\dfrac{2}{x-1} = 1$ **b.** $\dfrac{2}{x-2} = 2$ **c.** $\dfrac{-x-1}{x-3} = x+1$

d. $\dfrac{2}{x-1} = x$ **e.** $\dfrac{1}{x} = \dfrac{-1}{x-2}$ **f.** $\dfrac{1}{x} = x^2$

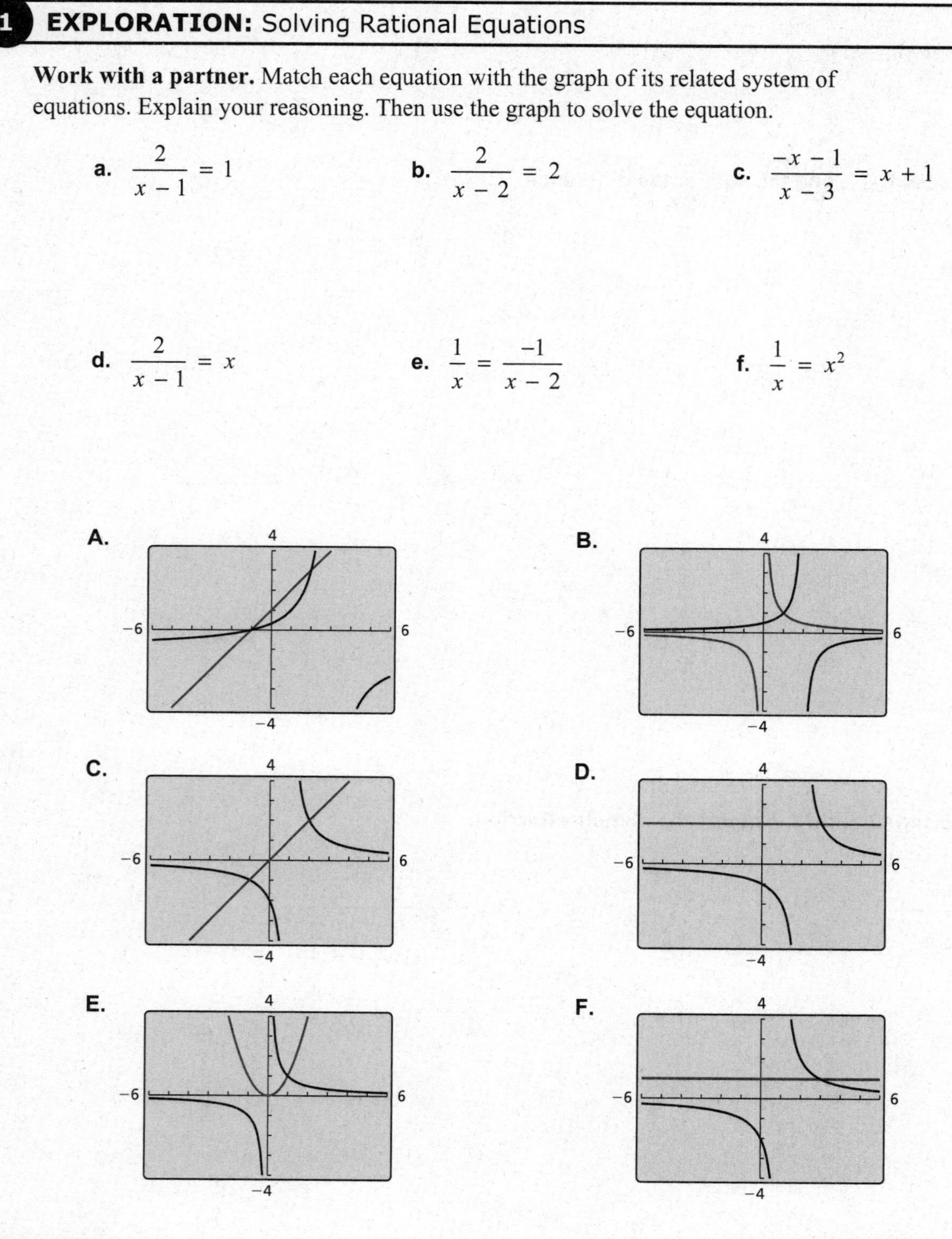

2 EXPLORATION: Solving Rational Equations

Go to *BigIdeasMath.com* for an interactive tool to investigate this exploration.

Work with a partner. Look back at the equations in Explorations 1(d) and 1(e). Suppose you want a more accurate way to solve the equations than using a graphical approach.

a. Show how you could use a *numerical approach* by creating a table. For instance, you might use a spreadsheet to solve the equations.

b. Show how you could use an *analytical approach*. For instance, you might use the method you used to solve proportions.

Communicate Your Answer

3. How can you solve a rational equation?

4. Use the method in either Exploration 1 or 2 to solve each equation.

a. $\frac{x+1}{x-1} = \frac{x-1}{x+1}$ **b.** $\frac{1}{x+1} = \frac{1}{x^2+1}$ **c.** $\frac{1}{x^2-1} = \frac{1}{x-1}$

Name ____ Date __________

Notetaking with Vocabulary
For use after Lesson 7.5

In your own words, write the meaning of each vocabulary term.

cross multiplying

Notes:

Extra Practice

In Exercises 1–4, solve the equation by cross multiplying. Check your solution(s).

1. $\dfrac{2}{x+8} = \dfrac{5}{2x-7}$

2. $\dfrac{x}{x+1} = \dfrac{-4}{x}$

3. $\dfrac{x+1}{x-3} = \dfrac{x+2}{x-6}$

4. $\dfrac{-2}{x-3} = \dfrac{x+9}{x+21}$

In Exercises 5–12, solve the equation by using the LCD. Check your solution(s).

5. $\dfrac{4}{7} - \dfrac{1}{x} = 6$

6. $\dfrac{3}{x+1} + \dfrac{4}{x+2} = \dfrac{15}{x+2}$

7. $\dfrac{12}{x+4} - \dfrac{7}{x} = \dfrac{22}{x^2+4x}$

8. $3 - \dfrac{18}{x-1} = -\dfrac{12}{x}$

7.5 Notetaking with Vocabulary (continued)

9. $\dfrac{2}{x-5} + \dfrac{3}{x} = \dfrac{10}{x^2 - 5x}$

10. $\dfrac{x+6}{x-4} - \dfrac{30}{x^2 - 5x + 4} = \dfrac{3}{x-1}$

11. $\dfrac{x}{x-5} + \dfrac{2}{x+2} = \dfrac{11}{x^2 - 3x - 10}$

12. $\dfrac{x-2}{x-4} - \dfrac{2}{x-1} = \dfrac{12}{x^2 - 5x + 4}$

In Exercises 13 and 14, determine whether the inverse of *f* is a function. Then find the inverse.

13. $f(x) = \dfrac{8}{x-3}$

14. $f(x) = \dfrac{12}{x} + 9$

15. You can complete the yard work at your friend's home in 5 hours. Working together, you and your friend can complete the yard work in 3 hours. How long would it take your friend to complete the yard work when working alone?

Let t be the time (in hours) your friend would take to complete the yard work when working alone.

	Work Rate	Time	Work Done
You	$\dfrac{1 \text{ yard}}{5 \text{ hours}}$	3 hours	
Friend		3 hours	

Name______________________________ Date__________

Maintaining Mathematical Proficiency

Complete the table to evaluate the function.

1. $y = 4 + 2^x$

x	y
1	
2	
3	

2. $y = 2x^3 - 5$

x	y
0	
1	
2	

3. $y = -2x + 10$

x	y
−4	
−2	
0	

Solve the equation. Check your solution(s).

4. $50 = 4 + 2x$

5. $\frac{1}{3} = 3\left(\frac{1}{3}\right)^x$

6. $45 = 5(2x - 1)$

7. $3^x + 12 = 93$

8. $\frac{2}{5}x + 8 = 2$

9. $\frac{32}{125} = 25\left(\frac{2}{5}\right)^x$

Name ________________________________ Date __________

8.1 Defining and Using Sequences and Series

For use with Exploration 8.1

Essential Question How can you write a rule for the *n*th term of a sequence?

A **sequence** is an ordered list of numbers. There can be a limited number or an infinite number of *terms* of a sequence.

$a_1, a_2, a_3, a_4, \ldots, a_n, \ldots$ Terms of a sequence

Here is an example.

$1, 4, 7, 10, \ldots, 3n - 2, \ldots$

EXPLORATION: Writing Rules for Sequences

Work with a partner. Match each sequence with its graph on the next page. The horizontal axes represent *n*, the position of each term in the sequence. Then write a rule for the *n*th term of the sequence, and use the rule to find a_{10}.

a. 1, 2.5, 4, 5.5, 7, ...

b. 8, 6.5, 5, 3.5, 2, ...

c. $\frac{1}{4}, \frac{4}{4}, \frac{9}{4}, \frac{16}{4}, \frac{25}{4}, \ldots$

d. $\frac{25}{4}, \frac{16}{4}, \frac{9}{4}, \frac{4}{4}, \frac{1}{4}, \ldots$

e. $\frac{1}{2}, 1, 2, 4, 8, \ldots$

f. $8, 4, 2, 1, \frac{1}{2}, \ldots$

Name____ Date__________

1 EXPLORATION: Writing Rules for Sequences (continued)

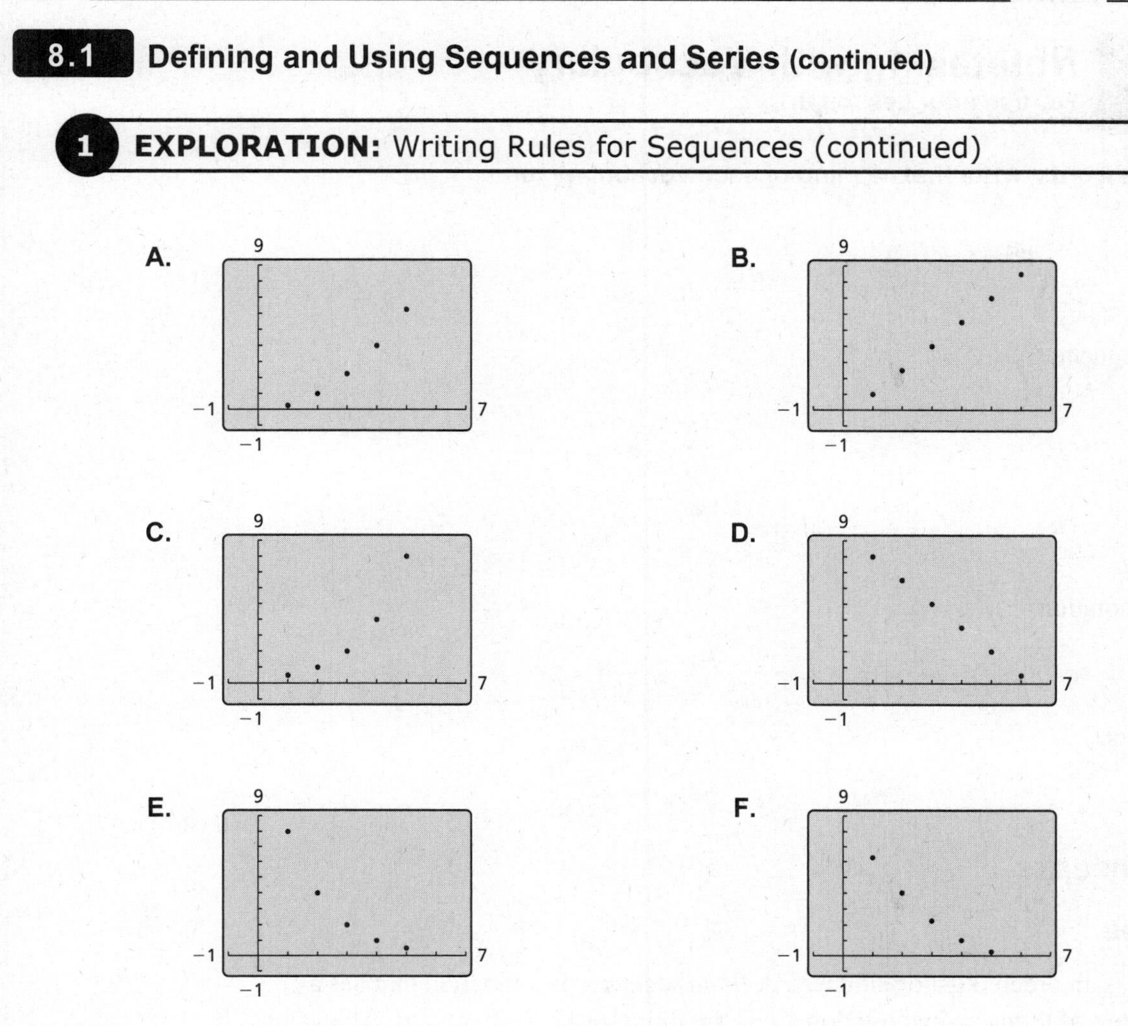

Communicate Your Answer

2. How can you write a rule for the *n*th term of a sequence?

3. What do you notice about the relationship between the terms in (a) an arithmetic sequence and (b) a geometric sequence? Justify your answers.

Name ______________________________ Date __________

8.1 Notetaking with Vocabulary

For use after Lesson 8.1

In your own words, write the meaning of each vocabulary term.

sequence

terms of a sequence

series

summation notation

sigma notation

Core Concepts

Sequences

A **sequence** is an ordered list of numbers. A *finite sequence* is a function that has a limited number of terms and whose domain is the finite set $\{1, 2, 3, \ldots, n\}$. The values in the range are called the **terms** of the sequence.

Domain:	1	2	3	4	...	n	Relative position of each term
	↓	↓	↓	↓		↓	
Range:	a_1	a_2	a_3	a_4	...	a_n	Terms of the sequence

An *infinite sequence* is a function that continues without stopping and whose domain is the set of positive integers. Here are examples of a finite sequence and an infinite sequence.

Finite sequence: 2, 4, 6, 8 **Infinite sequence:** 2, 4, 6, 8, ...

A sequence can be specified by an equation, or *rule*. For example, both sequences above can be described by the rule $a_n = 2n$ or $f(n) = 2n$.

Notes:

Series and Summation Notation

When the terms of a sequence are added together, the resulting expression is a **series**. A series can be finite or infinite.

Finite series: $2 + 4 + 6 + 8$

Infinite series: $2 + 4 + 6 + 8 + \cdots$

You can use **summation notation** to write a series. For example, the two series above can be written in summation notation as follows:

Finite series: $2 + 4 + 6 + 8 = \sum_{i=1}^{4} 2i$

Infinite series: $2 + 4 + 6 + 8 + \cdots = \sum_{i=1}^{\infty} 2i$

For both series, the *index of summation* is i and the *lower limit of summation* is 1. The *upper limit of summation* is 4 for the finite series and ∞ (infinity) for the infinite series. Summation notation is also called **sigma notation** because it uses the uppercase Greek letter *sigma*, written $\sum$.

Notes:

Formulas for Special Series

Sum of *n* terms of 1: $\sum_{i=1}^{n} 1 = n$

Sum of first *n* positive integers: $\sum_{i=1}^{n} i = \frac{n(n+1)}{2}$

Sum of squares of first *n* positive integers: $\sum_{i=1}^{n} i^2 = \frac{n(n+1)(2n+1)}{6}$

Notes:

Extra Practice

In Exercises 1 and 2, write the first six terms of the sequence.

1. $a_n = n^3 - 1$

2. $f(n) = (-2)^{n-1}$

In Exercises 3 and 4, describe the pattern, write the next term, and write a rule for the *n*th term of the sequence.

3. $-3, -1, 1, 3, \ldots$

4. $\frac{2}{5}, \frac{4}{5}, \frac{6}{5}, \frac{8}{5}, \ldots$

5. Write the series $-1 + 4 - 9 + 16 - 25 + \cdots$ using summation notation.

In Exercises 6 and 7, find the sum.

6. $\sum_{n=2}^{5} \frac{n}{n-1}$

7. $\sum_{i=1}^{18} i^2$

Name____ Date__________

8.2 Analyzing Arithmetic Sequences and Series

For use with Exploration 8.2

Essential Question How can you recognize an arithmetic sequence from its graph?

In an **arithmetic sequence**, the difference of consecutive terms, called the *common difference*, is constant. For example, in the arithmetic sequence 1, 4, 7, 10, . . . , the common difference is 3.

1 EXPLORATION: Recognizing Graphs of Arithmetic Sequences

Go to *BigIdeasMath.com* for an interactive tool to investigate this exploration.

Work with a partner. Determine whether each graph shows an arithmetic sequence. If it does, then write a rule for the *n*th term of the sequence, and use a spreadsheet to find the sum of the first 20 terms. What do you notice about the graph of an arithmetic sequence?

a.

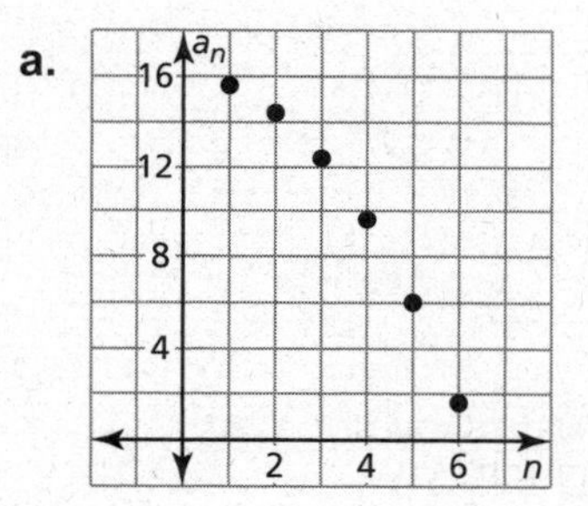

b.

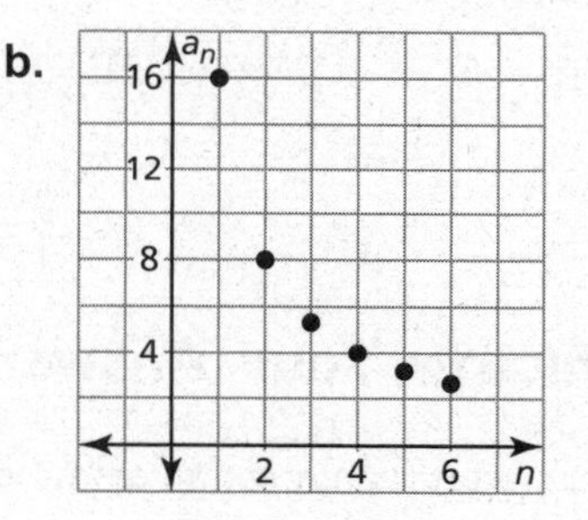

c.

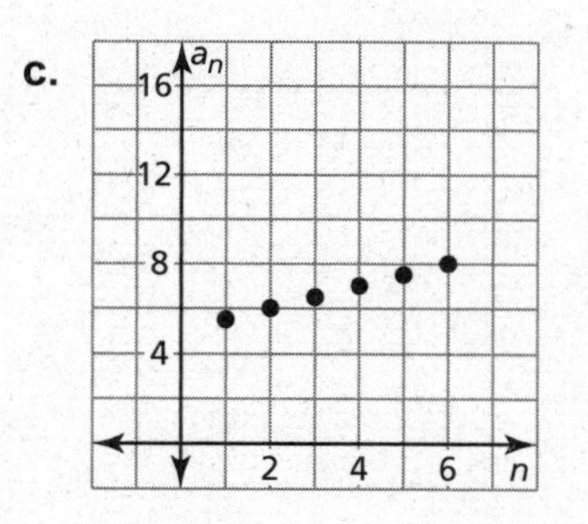

d.

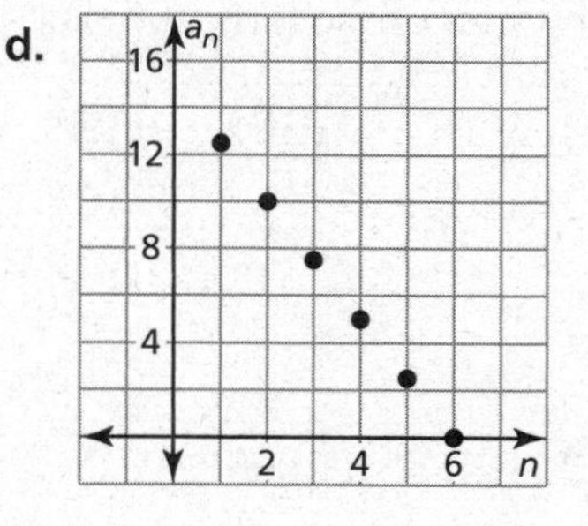

2 EXPLORATION: Finding the Sum of an Arithmetic Sequence

Work with a partner. A teacher of German mathematician Carl Friedrich Gauss (1777–1855) asked him to find the sum of all the whole numbers from 1 through 100. To the astonishment of his teacher, Gauss came up with the answer after only a few moments. Here is what Gauss did:

$$\begin{array}{rrrrr} 1\ + & 2\ + & 3\ + & \cdots\ + & 100 \\ 100\ + & 99\ + & 98\ + & \cdots\ + & 1 \\ \hline 101\ + & 101\ + & 101\ + & \cdots\ + & 101 \end{array} \qquad \frac{100 \times 101}{2} = 5050$$

Explain Gauss's thought process. Then write a formula for the sum S_n of the first n terms of an arithmetic sequence. Verify your formula by finding the sums of the first 20 terms of the arithmetic sequences in Exploration 1. Compare your answers to those you obtained using a spreadsheet.

Communicate Your Answer

3. How can you recognize an arithmetic sequence from its graph?

4. Find the sum of the terms of each arithmetic sequence.

a. 1, 4, 7, 10, . . . , 301

b. 1, 2, 3, 4, . . . , 1000

c. 2, 4, 6, 8, . . . , 800

Name____ Date__________

8.2 Notetaking with Vocabulary

For use after Lesson 8.2

In your own words, write the meaning of each vocabulary term.

arithmetic sequence

common difference

arithmetic series

Core Concepts

Rule for an Arithmetic Sequence

Algebra The nth term of an arithmetic sequence with first term a_1 and common difference d is given by:

$$a_n = a_1 + (n - 1)d$$

Example The nth term of an arithmetic sequence with a first term of 3 and a common difference of 2 is given by:

$$a_n = 3 + (n - 1)2, \text{ or } a_n = 2n + 1$$

Notes:

The Sum of a Finite Arithmetic Series

The sum of the first n terms of an arithmetic series is $S_n = n\left(\frac{a_1 + a_n}{2}\right)$.

In words, S_n is the mean of the first and nth terms, multiplied by the number of terms.

Notes:

Name ______________________________ Date __________

Extra Practice

In Exercises 1–4, tell whether the sequence is arithmetic. Explain your reasoning.

1. $1, 4, 7, 12, 17, \ldots$

2. $26, 23, 20, 17, 14, \ldots$

3. $0.3, 0.5, 0.7, 0.9, 1.1, \ldots$

4. $\frac{1}{2}, \frac{1}{4}, \frac{1}{6}, \frac{1}{8}, \frac{1}{10}, \ldots$

In Exercises 5–8, write a rule for the *n*th term of the sequence. Then find a_{20}.

5. $3, 9, 15, 21, \ldots$

6. $8, 3, -2, -7, \ldots$

7. $-1, -\frac{1}{2}, 0, \frac{1}{2}, \ldots$

8. $0.7, 0.2, -0.3, -0.8, \ldots$

9. Write a rule for the nth term of the sequence where $a_{12} = -13$ and $d = -2$. Then graph the first six terms of the sequence.

In Exercises 10 and 11, write a rule for the *n*th term of the sequence.

10. $a_8 = 59, a_{13} = 99$

11. $a_{18} = -5, a_{27} = -8$

12. Find the sum $\sum_{i=1}^{22} (5 - 2i)$.

Name ____ Date __________

8.3 Analyzing Geometric Sequences and Series

For use with Exploration 8.3

Essential Question How can you recognize a geometric sequence from its graph?

In a **geometric sequence**, the ratio of any term to the previous term, called the *common ratio*, is constant. For example, in the geometric sequence 1, 2, 4, 8, . . . , the common ratio is 2.

1 EXPLORATION: Recognizing Graphs of Geometric Sequences

Go to *BigIdeasMath.com* for an interactive tool to investigate this exploration.

Work with a partner. Determine whether each graph shows a geometric sequence. If it does, then write a rule for the nth term of the sequence and use a spreadsheet to find the sum of the first 20 terms. What do you notice about the graph of a geometric sequence?

a.

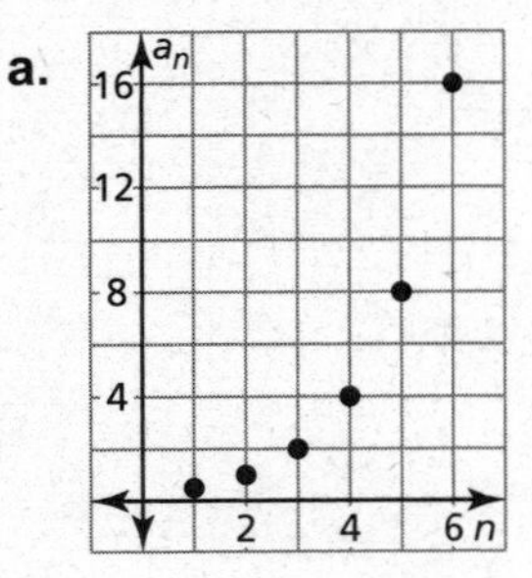

b.

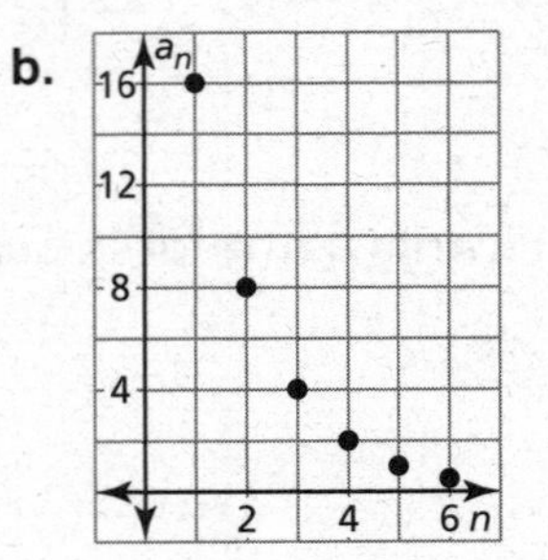

c.

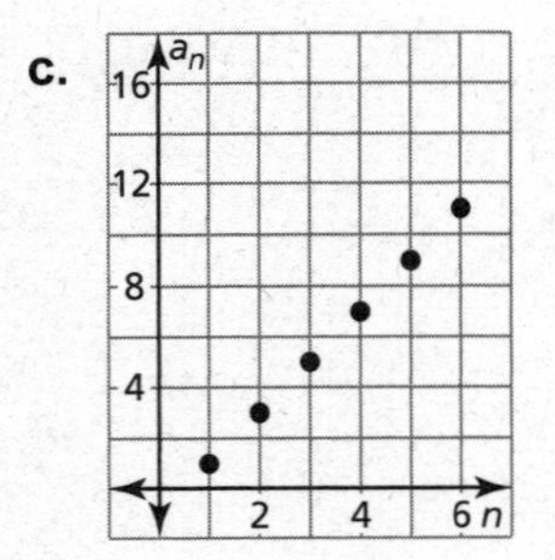

d.

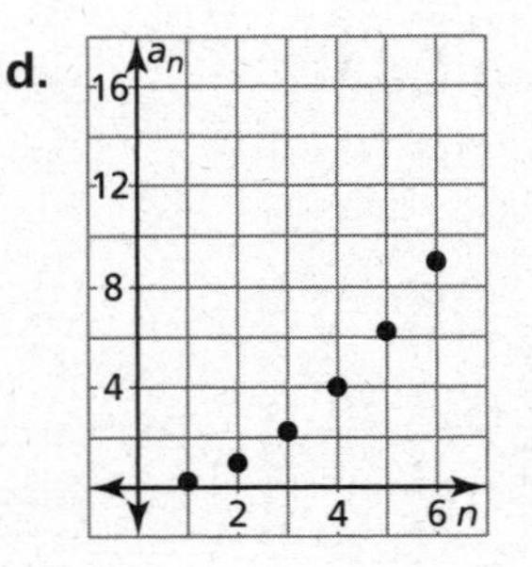

2 EXPLORATION: Finding the Sum of a Geometric Sequence

Work with a partner. You can write the *n*th term of a geometric sequence with first term a_1 and common ratio r as

$$a_n = a_1 r^{n-1}.$$

So, you can write the sum S_n of the first n terms of a geometric sequence as

$$S_n = a_1 + a_1 r + a_1 r^2 + a_1 r^3 + \cdots + a_1 r^{n-1}.$$

Rewrite this formula by finding the difference $S_n - rS_n$ and solve for S_n. Then verify your rewritten formula by finding the sums of the first 20 terms of the geometric sequences in Exploration 1. Compare your answers to those you obtained using a spreadsheet.

Communicate Your Answer

3. How can you recognize a geometric sequence from its graph?

4. Find the sum of the terms of each geometric sequence.

a. $1, 2, 4, 8, \ldots, 8192$

b. $0.1, 0.01, 0.001, 0.0001, \ldots, 10^{-10}$

Name ____ Date __________

8.3 Notetaking with Vocabulary

For use after Lesson 8.3

In your own words, write the meaning of each vocabulary term.

geometric sequence

common ratio

geometric series

Core Concepts

Rule for a Geometric Sequence

Algebra The nth term of a geometric sequence with first term a_1 and common ratio r is given by:

$$a_n = a_1 r^{n-1}$$

Example The nth term of a geometric sequence with a first term of 2 and a common ratio of 3 is given by:

$$a_n = 2(3)^{n-1}$$

Notes:

The Sum of a Finite Geometric Series

The sum of the first n terms of a geometric series with common ratio $r \neq 1$ is

$$S_n = a_1\left(\frac{1 - r^n}{1 - r}\right).$$

Notes:

8.3 Notetaking with Vocabulary (continued)

Extra Practice

In Exercises 1–4, tell whether the sequence is geometric. Explain your reasoning.

1. $4, 12, 36, 108, 324, \ldots$

2. $45, 40, 35, 30, 25, \ldots$

3. $1.3, 7.8, 46.8, 280.8, 1684.8, \ldots$

4. $\frac{3}{2}, -\frac{3}{4}, \frac{3}{8}, -\frac{3}{16}, \frac{3}{32}, \ldots$

In Exercises 5–8, write a rule for the *n*th term of the sequence. Then find a_6.

5. $6, 18, 54, 162, \ldots$

6. $3, -6, 12, -24, \ldots$

7. $1, \frac{5}{2}, \frac{25}{4}, \frac{125}{8}, \ldots$

8. $-2.4, -16.8, -117.6, -823.2, \ldots$

Name ______________________________ Date __________

8.3 Notetaking with Vocabulary (continued)

9. Write a rule for the *n*th term where $a_8 = 384$ and $r = 2$. Then graph the first six terms of the sequence.

In Exercises 10 and 11, write a rule for the *n*th term of the geometric sequence.

10. $a_3 = 54, a_6 = 1458$

11. $a_2 = -2, a_5 = \frac{2}{125}$

12. Find the sum $\sum_{i=0}^{10} 3\left(\frac{3}{2}\right)^{i-1}$.

Name________________________________ Date__________

8.4 Finding Sums of Infinite Geometric Series

For use with Exploration 8.4

Essential Question How can you find the sum of an infinite geometric series?

1 EXPLORATION: Finding Sums of Infinite Geometric Series

Go to *BigIdeasMath.com* for an interactive tool to investigate this exploration.

Work with a partner. Enter each geometric series in a spreadsheet. Then use the spreadsheet to determine whether the infinite geometric series has a finite sum. If it does, find the sum. Explain your reasoning. (The figure shows a partially completed spreadsheet for part (a).)

a. $1 + \frac{1}{2} + \frac{1}{4} + \frac{1}{8} + \frac{1}{16} + \cdots$

	A	B
1	1	1
2	2	0.5
3	3	0.25
4	4	0.125
5	5	0.0625
6	6	0.03125
7	7	
8	8	
9	9	
10	10	
11	11	
12	12	
13	13	
14	14	
15	15	
16	Sum	

b. $1 + \frac{1}{3} + \frac{1}{9} + \frac{1}{27} + \frac{1}{81} + \cdots$

c. $1 + \frac{3}{2} + \frac{9}{4} + \frac{27}{8} + \frac{81}{16} + \cdots$

d. $1 + \frac{5}{4} + \frac{25}{16} + \frac{125}{64} + \frac{625}{256} + \cdots$

e. $1 + \frac{4}{5} + \frac{16}{25} + \frac{64}{125} + \frac{256}{625} + \cdots$

f. $1 + \frac{9}{10} + \frac{81}{100} + \frac{729}{1000} + \frac{6561}{10{,}000} + \cdots$

Name ____ Date __________

8.4 Finding Sums of Infinite Geometric Series (continued)

2 EXPLORATION: Writing a Conjecture

Work with a partner. Look back at the infinite geometric series in Exploration 1. Write a conjecture about how you can determine whether the infinite geometric series

$$a_1 + a_1r + a_1r^2 + a_1r^3 + \cdots$$

has a finite sum.

3 EXPLORATION: Writing a Formula

Work with a partner. In Lesson 8.3, you learned that the sum of the first n terms of a geometric series with first term a_1 and common ratio $r \neq 1$ is

$$S_n = a_1\left(\frac{1 - r^n}{1 - r}\right).$$

When an infinite geometric series has a finite sum, what happens to r^n as n increases? Explain your reasoning. Write a formula to find the sum of an infinite geometric series. Then verify your formula by checking the sums you obtained in Exploration 1.

Communicate Your Answer

4. How can you find the sum of an infinite geometric series?

5. Find the sum of each infinite geometric series, if it exists.

a. $1 + 0.1 + 0.01 + 0.001 + 0.0001 + \cdots$

b. $2 + \frac{4}{3} + \frac{8}{9} + \frac{16}{27} + \frac{32}{81} + \cdots$

Name____ Date__________

8.4 Notetaking with Vocabulary
For use after Lesson 8.4

In your own words, write the meaning of each vocabulary term.

partial sum

Core Concepts

The Sum of an Infinite Geometric Series

The sum of an infinite geometric series with first term a_1 and common ratio r is given by

$$S = \frac{a_1}{1 - r}$$

provided $|r| < 1$. If $|r| \geq 1$, then the series has no sum.

Notes:

Extra Practice

In Exercises 1 and 2, consider the infinite geometric series. Find and graph the partial sums S_n for $n = 1, 2, 3, 4$, and 5. Then describe what happens to S_n as n increases.

1. $\frac{1}{4} + \frac{1}{6} + \frac{1}{9} + \frac{2}{27} + \frac{4}{81} + \cdots$

2. $3 + \frac{3}{5} + \frac{3}{25} + \frac{3}{125} + \frac{3}{625} + \cdots$

Name____ Date__________

8.4 Notetaking with Vocabulary (continued)

In Exercises 3–6, find the sum of the infinite geometric series, if it exists.

3. $\sum_{n=1}^{\infty} 6\left(\frac{3}{5}\right)^{n-1}$

4. $\sum_{i=1}^{\infty} \frac{10}{3}\left(\frac{5}{2}\right)^{i-1}$

5. $5 + \frac{5}{3} + \frac{5}{9} + \frac{5}{27} + \cdots$

6. $\frac{1}{2} - \frac{1}{4} + \frac{1}{8} - \frac{1}{16} + \cdots$

7. A child pushes a tumbler toy and lets it swing freely. On the first swing, the toy travels 30 centimeters. On each successive swing, the toy travels 75% of the distance of the previous swing. What is the total distance the toy swings?

8. Write 0.121212… as a fraction in simplest form.

Name ______________________________ Date __________

8.5 Using Recursive Rules with Sequences

For use with Exploration 8.5

Essential Question How can you define a sequence recursively?

A **recursive rule** gives the beginning term(s) of a sequence and a *recursive equation* that tells how a_n is related to one or more preceding terms.

1 EXPLORATION: Evaluating a Recursive Rule

Go to *BigIdeasMath.com* for an interactive tool to investigate this exploration.

Work with a partner. Use each recursive rule and a spreadsheet to write the first six terms of the sequence. Classify the sequence as arithmetic, geometric, or neither. Explain your reasoning. (The figure shows a partially completed spreadsheet for part (a).)

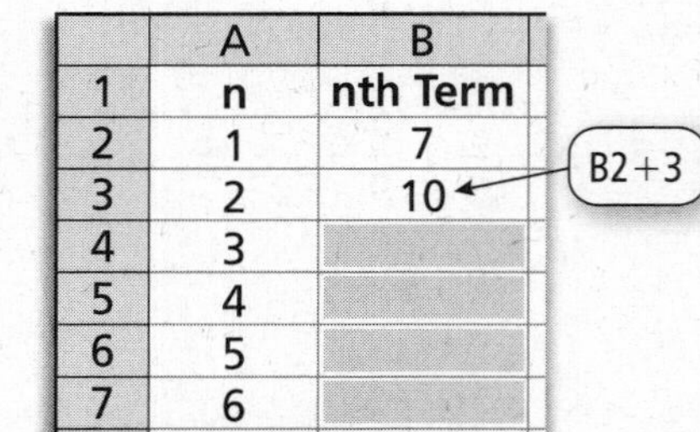

	A	B
1	n	nth Term
2	1	7
3	2	10
4	3	
5	4	
6	5	
7	6	

a. $a_1 = 7, a_n = a_{n-1} + 3$

b. $a_1 = 5, a_n = a_{n-1} - 2$

c. $a_1 = 1, a_n = 2a_{n-1}$

d. $a_1 = 1, a_n = \frac{1}{2}(a_{n-1})^2$

e. $a_1 = 3, a_n = a_{n-1} + 1$

f. $a_1 = 4, a_n = \frac{1}{2}a_{n-1} - 1$

g. $a_1 = 4, a_n = \frac{1}{2}a_{n-1}$

h. $a_1 = 4, a_2 = 5, a_n = a_{n-1} + a_{n-2}$

Name____ Date__________

8.5 Using Recursive Rules with Sequences (continued)

2 EXPLORATION: Writing a Recursive Rule

Work with a partner. Write a recursive rule for the sequence. Explain your reasoning.

a. $3, 6, 9, 12, 15, 18, \ldots$

b. $18, 14, 10, 6, 2, -2, \ldots$

c. $3, 6, 12, 24, 48, 96, \ldots$

d. $128, 64, 32, 16, 8, 4, \ldots$

e. $5, 5, 5, 5, 5, 5, \ldots$

f. $1, 1, 2, 3, 5, 8, \ldots$

3 EXPLORATION: Writing a Recursive Rule

Work with a partner. Write a recursive rule for the sequence whose graph is shown.

a.

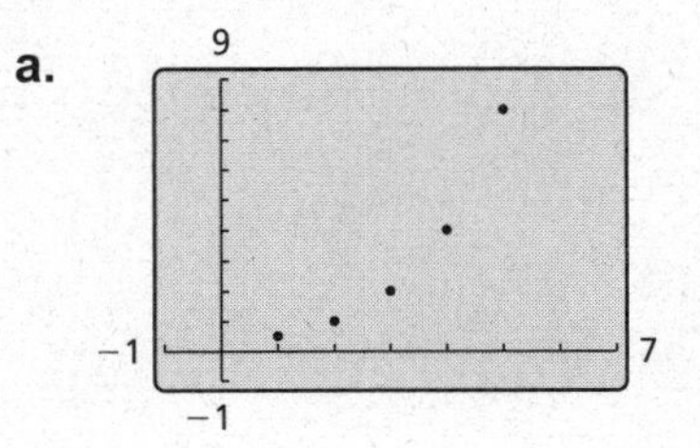

b.

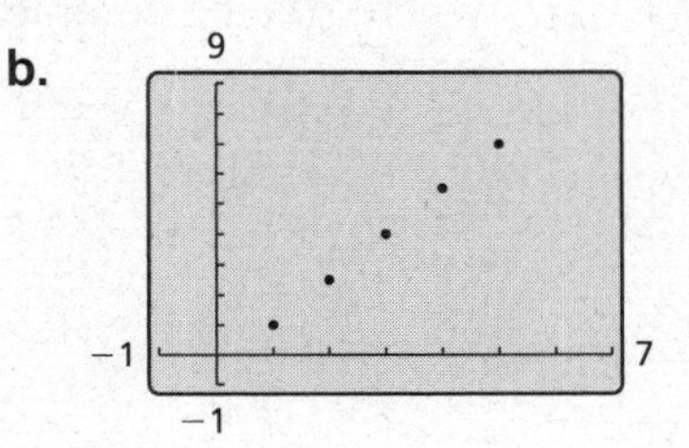

Communicate Your Answer

4. How can you define a sequence recursively?

5. Write a recursive rule that is different from those in Explorations 1–3. Write the first six terms of the sequence. Then graph the sequence and classify it as arithmetic, geometric, or neither.

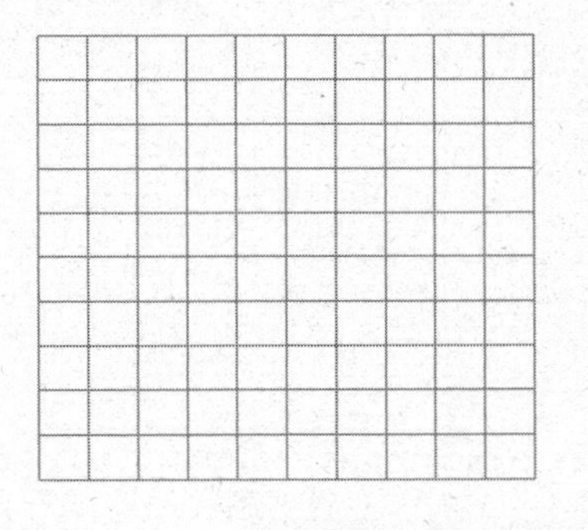

Name ______________________________ Date __________

8.5 Notetaking with Vocabulary
For use after Lesson 8.5

In your own words, write the meaning of each vocabulary term.

explicit rule

recursive rule

Core Concepts

Recursive Equations for Arithmetic and Geometric Sequences

Arithmetic Sequence

$a_n = a_{n-1} + d$, where d is the common difference

Geometric Sequence

$a_n = r \bullet a_{n-1}$, where r is the common ratio

Notes:

Name____ Date__________

8.5 Notetaking with Vocabulary (continued)

Extra Practice

In Exercises 1 and 2, write the first six terms of the sequence.

1. $a_1 = 2$

$a_n = a_{n-1} + 5$

2. $f(0) = 1$

$f(n) = 2f(n-1)$

In Exercises 3–6, write a recursive rule for the sequence.

3. $9, 12, 15, 18, 21, \ldots$

4. $50, 20, 8, \frac{16}{5}, \frac{32}{25}, \ldots$

5. $3, 4, 1, -3, -4, \ldots$

6. $1, 1, \frac{1}{3}, \frac{1}{4}, \frac{1}{15}, \ldots$

In Exercises 7–10, write a recursive rule for the sequence.

7. $a_n = 5 - 3n$

8. $a_n = 10(-2)^{n-1}$

9. $a_n = -1 + 8n$

10. $a_n = -3\left(\frac{3}{4}\right)^{n-1}$

In Exercises 11–14, write an explicit rule for each sequence.

11. $a_1 = -1, a_n = a_{n-1} + 7$

12. $a_1 = 24, a_n = 0.2a_{n-1}$

13. $a_1 = 1, a_n = a_{n-1} - 0.3$

14. $a_1 = -2, a_n = -5a_{n-1}$

Name______________________________ Date__________

Chapter 9 Maintaining Mathematical Proficiency

Order the expressions by value from least to greatest.

1. $|-8|, |1+4|, |3-7|, -|2|$

2. $|8-2|, |3-6|, |0|, \left|-\frac{3}{2}\right|$

3. $|-2^2|, |-9-1|, |3 \bullet (-3)|, |-2| + |-1| - |3|$

4. $|-5+15|, |3| - |3 \bullet 4|, |3-7|, -|3^3|$

Find the missing side length of the triangle.

5.

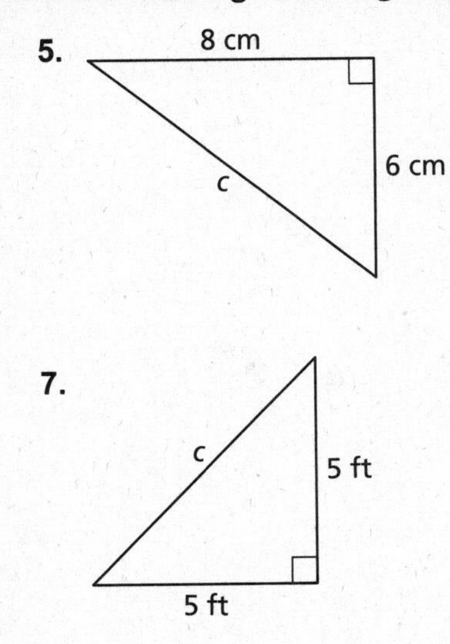

6.

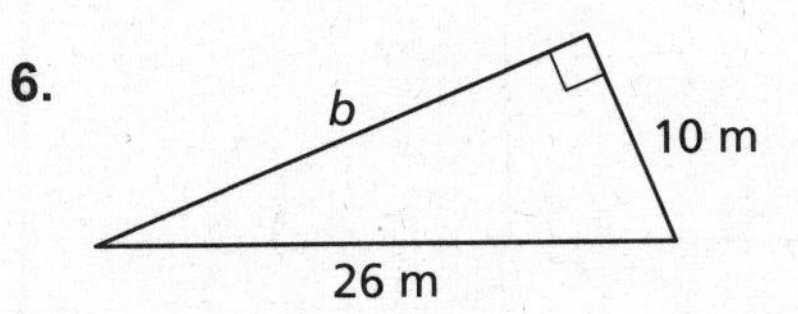

7.

8.

a

1.7 in.

0.8 in.

Name ______________________________ Date __________

9.1 Right Triangle Trigonometry

For use with Exploration 9.1

Essential Question How can you find a trigonometric function of an acute angle θ?

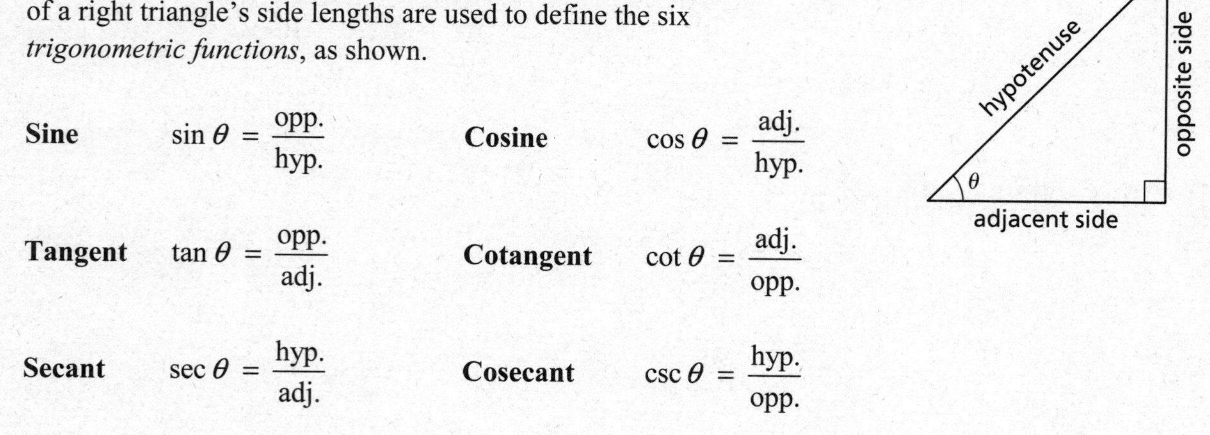

Consider one of the acute angles θ of a right triangle. Ratios of a right triangle's side lengths are used to define the six *trigonometric functions*, as shown.

Sine $\sin\theta = \dfrac{\text{opp.}}{\text{hyp.}}$ **Cosine** $\cos\theta = \dfrac{\text{adj.}}{\text{hyp.}}$

Tangent $\tan\theta = \dfrac{\text{opp.}}{\text{adj.}}$ **Cotangent** $\cot\theta = \dfrac{\text{adj.}}{\text{opp.}}$

Secant $\sec\theta = \dfrac{\text{hyp.}}{\text{adj.}}$ **Cosecant** $\csc\theta = \dfrac{\text{hyp.}}{\text{opp.}}$

1 EXPLORATION: Trigonometric Functions of Special Angles

Work with a partner. Find the exact values of the sine, cosine, and tangent functions for the angles 30°, 45°, and 60° in the right triangles shown.

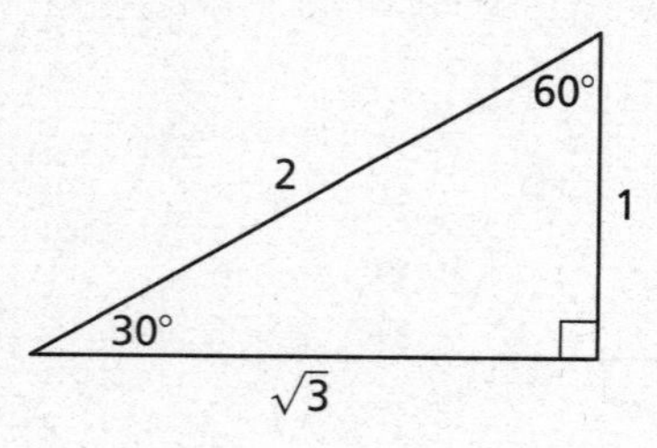

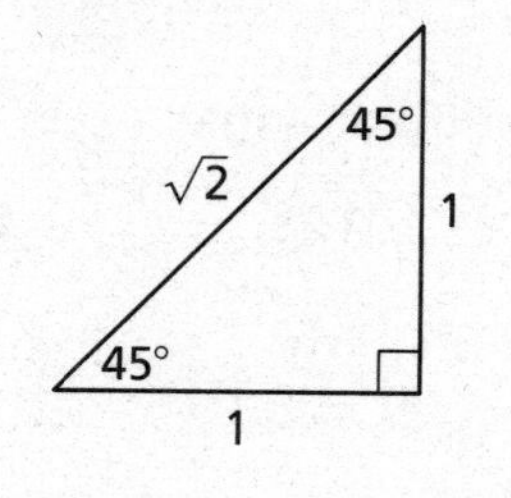

2 EXPLORATION: Exploring Trigonometric Identities

Work with a partner.

Use the definitions of the trigonometric functions to explain why each *trigonometric identity* is true.

a. $\sin \theta = \cos(90° - \theta)$

b. $\cos \theta = \sin(90° - \theta)$

c. $\sin \theta = \dfrac{1}{\csc \theta}$

d. $\tan \theta = \dfrac{1}{\cot \theta}$

Use the definitions of the trigonometric functions to complete each trigonometric identity.

e. $(\sin \theta)^2 + (\cos \theta)^2 =$ ________

f. $(\sec \theta)^2 - (\tan \theta)^2 =$ ________

Communicate Your Answer

3. How can you find a trigonometric function of an acute angle θ?

4. Use a calculator to find the lengths x and y of the legs of the right triangle shown.

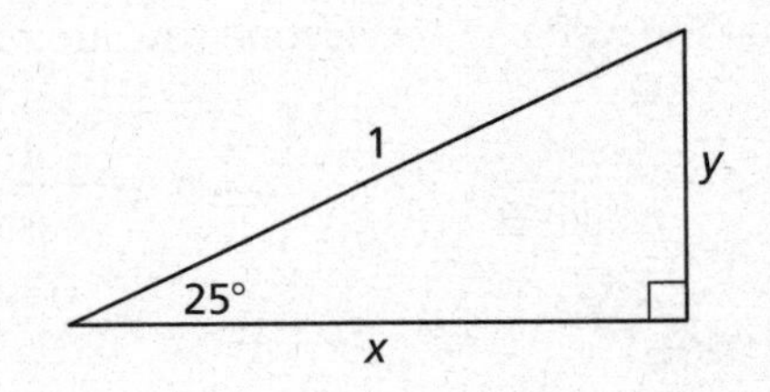

Name ______________________________ Date ________

9.1 Notetaking with Vocabulary
For use after Lesson 9.1

In your own words, write the meaning of each vocabulary term.

sine

cosine

tangent

cosecant

secant

cotangent

Core Concepts

Right Triangle Definitions of Trigonometric Functions

Let θ be an acute angle of a right triangle. The six trigonometric functions of θ are defined as shown.

$\sin \theta = \dfrac{\text{opposite}}{\text{hypotenuse}}$ $\cos \theta = \dfrac{\text{adjacent}}{\text{hypotenuse}}$ $\tan \theta = \dfrac{\text{opposite}}{\text{adjacent}}$

$\csc \theta = \dfrac{\text{hypotenuse}}{\text{opposite}}$ $\sec \theta = \dfrac{\text{hypotenuse}}{\text{adjacent}}$ $\cot \theta = \dfrac{\text{adjacent}}{\text{opposite}}$

The abbreviations *opp.*, *adj.*, and *hyp.* are often used to represent the side lengths of the right triangle. Note that the ratios in the second row are reciprocals of the ratios in the first row.

$\csc \theta = \dfrac{1}{\sin \theta}$ $\sec \theta = \dfrac{1}{\cos \theta}$ $\cot \theta = \dfrac{1}{\tan \theta}$

Notes:

Name____ Date__________

9.1 Notetaking with Vocabulary (continued)

Trigonometric Values for Special Angles

The table gives the values of the six trigonometric functions for the angles 30°, 45°, and 60°. You can obtain these values from the triangles shown.

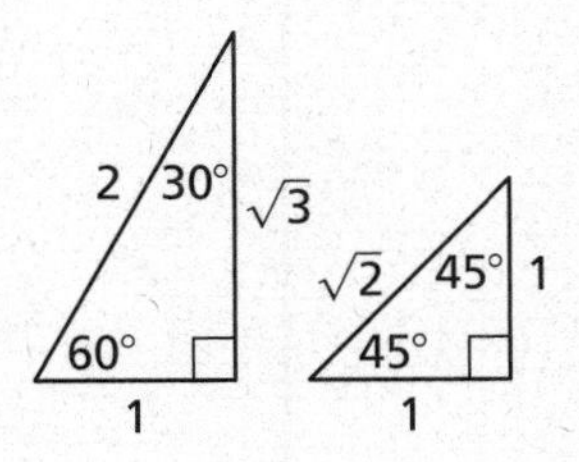

θ	$\sin\theta$	$\cos\theta$	$\tan\theta$	$\csc\theta$	$\sec\theta$	$\cot\theta$
30°	$\frac{1}{2}$	$\frac{\sqrt{3}}{2}$	$\frac{\sqrt{3}}{3}$	2	$\frac{2\sqrt{3}}{3}$	$\sqrt{3}$
45°	$\frac{\sqrt{2}}{2}$	$\frac{\sqrt{2}}{2}$	1	$\sqrt{2}$	$\sqrt{2}$	1
60°	$\frac{\sqrt{3}}{2}$	$\frac{1}{2}$	$\sqrt{3}$	$\frac{2\sqrt{3}}{3}$	2	$\frac{\sqrt{3}}{3}$

Notes:

Extra Practice

In Exercises 1 and 2, evaluate the six trigonometric functions of the angle θ.

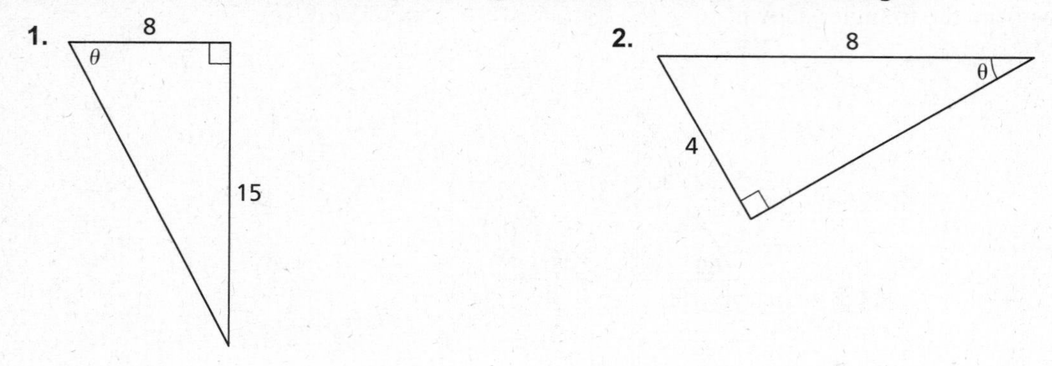

In Exercises 3 and 4, let θ be an acute angle of a right triangle. Evaluate the other five trigonometric functions of θ.

3. $\tan \theta = 1$

4. $\sin \theta = \frac{3}{19}$

In Exercises 5 and 6, find the value of x for the right triangle.

Name____ Date__________

9.2 Angles and Radian Measure

For use with Exploration 9.2

Essential Question How can you find the measure of an angle in radians?

Let the vertex of an angle be at the origin, with one side of the angle on the positive x-axis. The *radian measure* of the angle is a measure of the intercepted arc length on a circle of radius 1. To convert between degree and radian measure, use the fact that $\frac{\pi \text{ radians}}{180^\circ} = 1.$

1 EXPLORATION: Writing Radian Measures of Angles

Work with a partner. Write the radian measure of each angle with the given degree measure. Explain your reasoning.

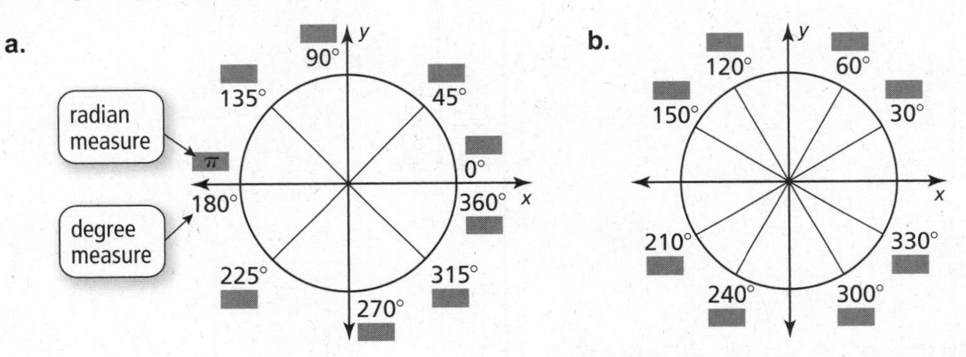

Name ______________________________ Date __________

2 EXPLORATION: Writing Degree Measures of Angles

Work with a partner. Write the degree measure of each angle with the given radian measure. Explain your reasoning.

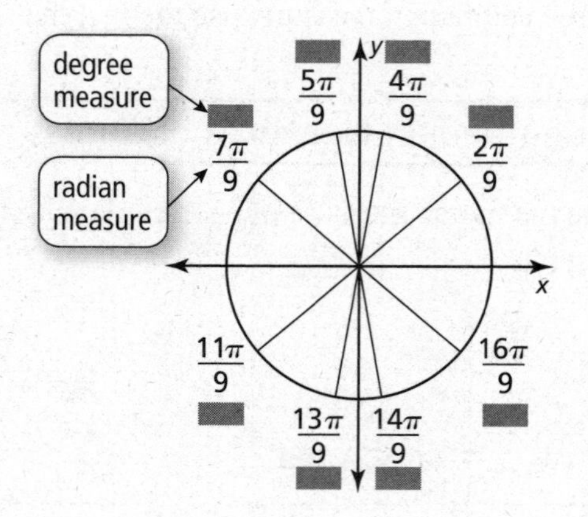

Communicate Your Answer

3. How can you find the measure of an angle in radians?

4. The figure shows an angle whose measure is 30 radians. What is the measure of the angle in degrees? How many times greater is 30 radians than 30 degrees? Justify your answers.

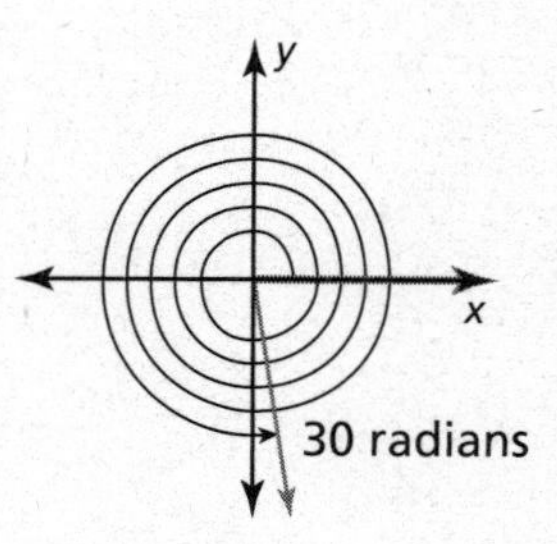

Name______________________________ Date__________

9.2 Notetaking with Vocabulary
For use after Lesson 9.2

In your own words, write the meaning of each vocabulary term.

initial side

terminal side

standard position

coterminal

radian

sector

central angle

Core Concepts

Angles in Standard Position

In a coordinate plane, an angle can be formed by fixing one ray, called the **initial side**, and rotating the other ray, called the **terminal side**, about the vertex.

An angle is in **standard position** when its vertex is at the origin and its initial side lies on the positive x-axis.

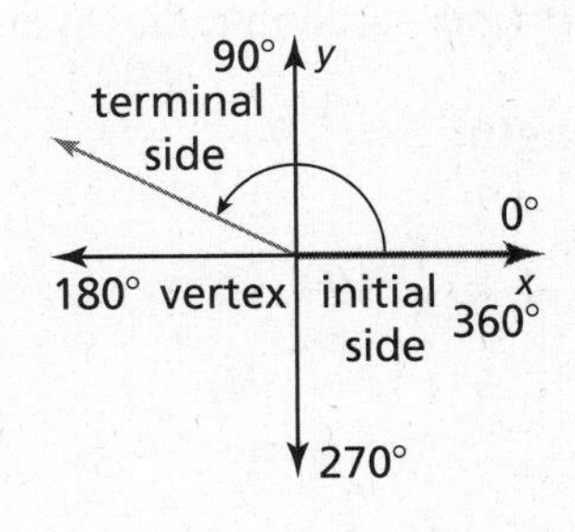

Notes:

Name ______________________________ Date __________

Converting Between Degrees and Radians

Degrees to radians

Multiply degree measure by

$$\frac{\pi \text{ radians}}{180^\circ}.$$

Radians to degrees

Multiply radian measure by

$$\frac{180^\circ}{\pi \text{ radians}}.$$

Degree and Radian Measures of Special Angles

The diagram shows equivalent degree and radian measures for special angles from 0° to 360° (0 radians to 2π radians).

You may find it helpful to memorize the equivalent degree and radian measures of special angles in the first quadrant and for $90^\circ = \frac{\pi}{2}$ radians. All other special angles shown are multiples of theses angles.

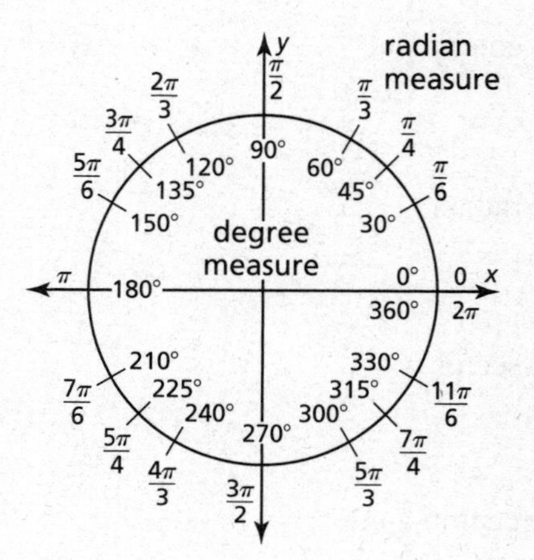

Arc Length and Area of a Sector

The arc length s and area A of a sector with radius r and central angle θ (measured in radians) are as follows.

Arc length: $s = r\theta$

Area: $A = \frac{1}{2}r^2\theta$

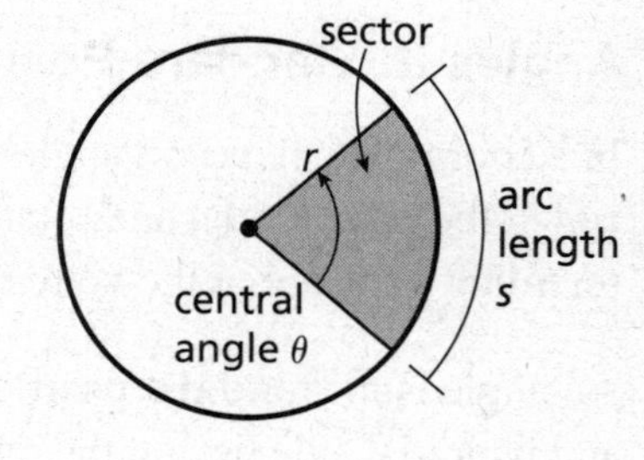

Notes:

Name____ Date__________

Extra Practice

In Exercises 1 and 2, draw an angle with the given measure in standard position.

1. 260° **2.** -750°

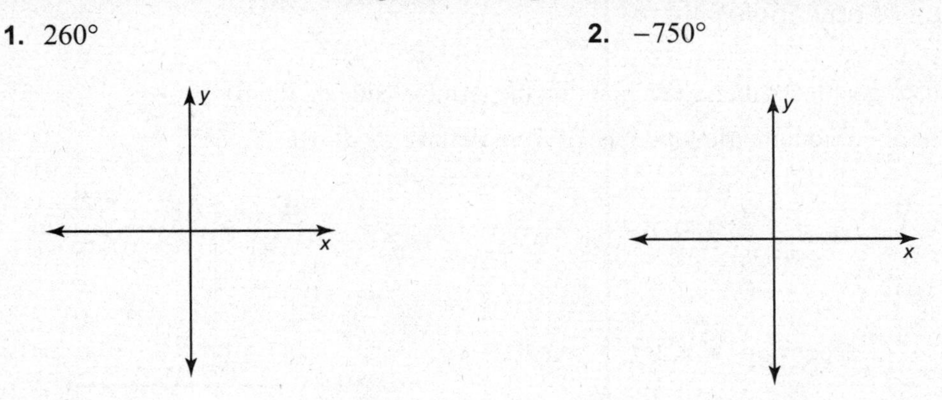

In Exercises 3–6, find one positive angle and one negative angle that are coterminal with the given angle.

3. 55°

4. -300°

5. 460°

6. -220°

In Exercises 7–10, convert the degree measure to radians or the radian measure to degrees.

7. 54°

8. -310°

9. $\dfrac{16\pi}{15}$

10. $-\dfrac{2\pi}{5}$

Name ______________________________ Date __________

9.3 Trigonometric Functions of Any Angle

For use with Exploration 9.3

Essential Question How can you use the unit circle to define the trigonometric functions of any angle?

Let θ be an angle in standard position with (x, y) a point on the terminal side of θ and $r = \sqrt{x^2 + y^2} \neq 0$. The six trigonometric functions of θ are defined as shown.

$\sin \theta = \dfrac{y}{r}$ $\qquad$ $\csc \theta = \dfrac{r}{y}, y \neq 0$

$\cos \theta = \dfrac{x}{r}$ $\qquad$ $\sec \theta = \dfrac{r}{x}, x \neq 0$

$\tan \theta = \dfrac{y}{x}, x \neq 0$ $\qquad$ $\cot \theta = \dfrac{x}{y}, y \neq 0$

1 EXPLORATION: Writing Trigonometric Functions

Work with a partner. Find the sine, cosine, and tangent of the angle θ in standard position whose terminal side intersects the unit circle at the point (x, y) shown.

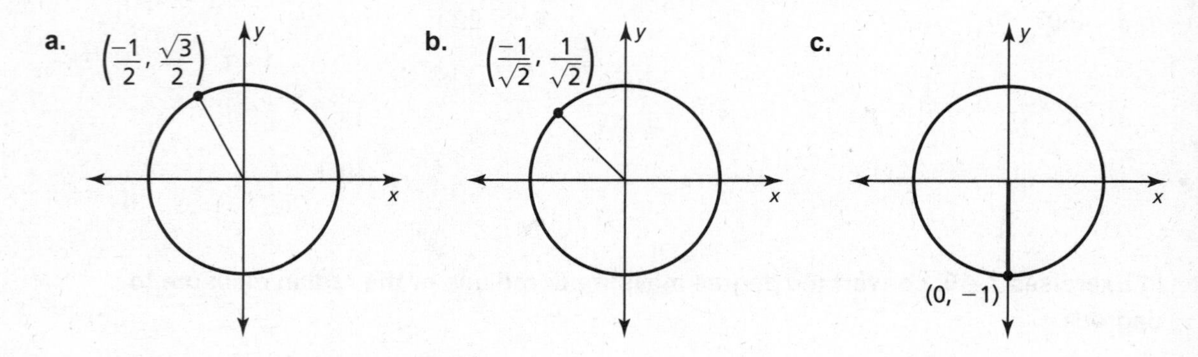

Name____________________ Date__________

1 EXPLORATION: Writing Trigonometric Functions (continued)

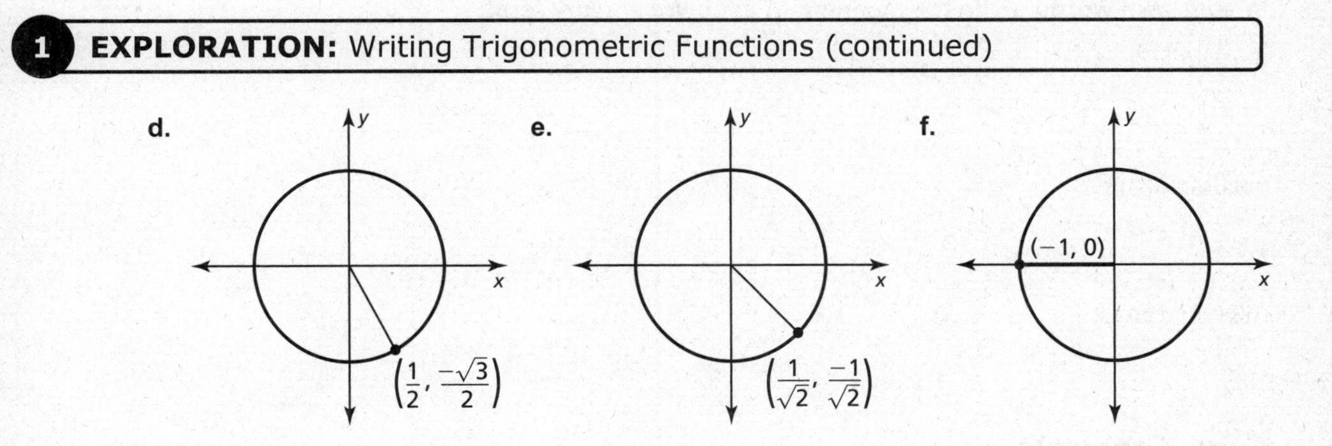

Communicate Your Answer

2. How can you use the unit circle to define the trigonometric functions of any angle?

3. For which angles are each function undefined? Explain your reasoning.

a. tangent

b. cotangent

c. secant

d. cosecant

Name ______________________________ Date __________

9.3 Notetaking with Vocabulary
For use after Lesson 9.3

In your own words, write the meaning of each vocabulary term.

unit circle

quadrantal angle

reference angle

Core Concepts

General Definitions of Trigonometric Functions

Let θ be an angle in standard position, and let (x, y) be the point where the terminal side of θ intersects the circle $x^2 + y^2 = r^2$. The six trigonometric functions of θ are defined as shown.

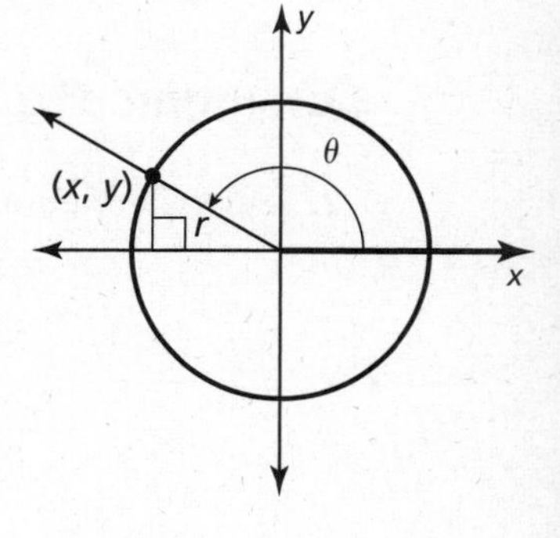

$\sin \theta = \dfrac{y}{r}$ $\qquad$ $\csc \theta = \dfrac{r}{y}, y \neq 0$

$\cos \theta = \dfrac{x}{r}$ $\qquad$ $\sec \theta = \dfrac{r}{x}, x \neq 0$

$\tan \theta = \dfrac{y}{x}, x \neq 0$ $\qquad$ $\cot \theta = \dfrac{x}{y}, y \neq 0$

These functions are sometimes called *circular functions*.

The Unit Circle

The circle $x^2 + y^2 = 1$, which has center $(0, 0)$ and radius 1, is called the **unit circle**. The values of $\sin \theta$ and $\cos \theta$ are simply the y-coordinate and x-coordinate, respectively, of the point where the terminal side of θ intersects the unit circle.

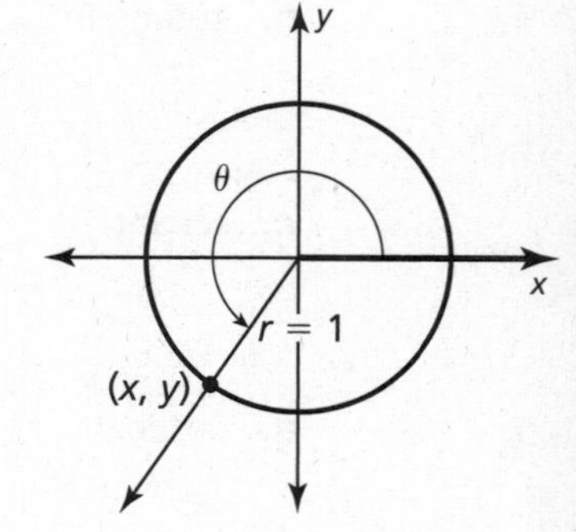

$\sin \theta = \dfrac{y}{r} = \dfrac{y}{1} = y$ $\qquad$ $\cos \theta = \dfrac{x}{r} = \dfrac{x}{1} = x$

Notes:

Name____ Date__________

9.3 Notetaking with Vocabulary (continued)

Reference Angle Relationships

Let θ be an angle in standard position. The **reference angle** for θ is the acute angle θ' formed by the terminal side of θ and the x-axis. The relationship between θ and θ' is shown below for nonquadrantal angles θ such that $90° < \theta < 360°$ or, in radians, $\frac{\pi}{2} < \theta < 2\pi$.

Quadrant II

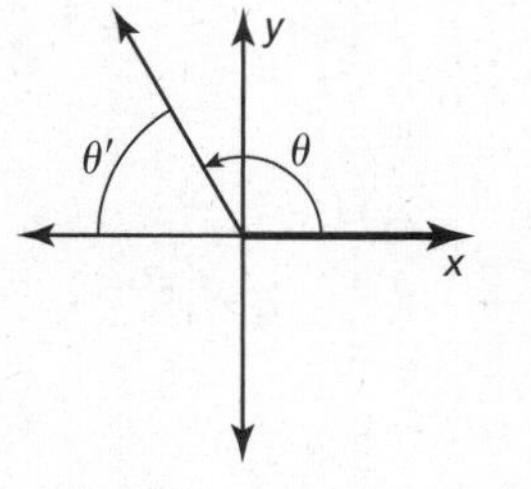

Degrees: $\theta' = 180° - \theta$

Radians: $\theta' = \pi - \theta$

Quadrant III

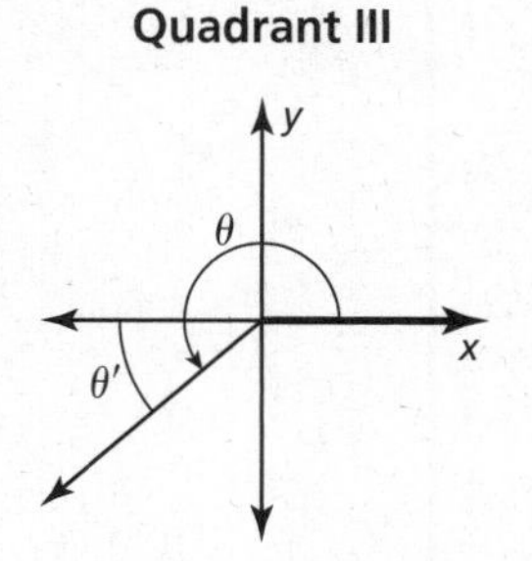

Degrees: $\theta' = \theta - 180°$

Radians: $\theta' = \theta - \pi$

Quadrant IV

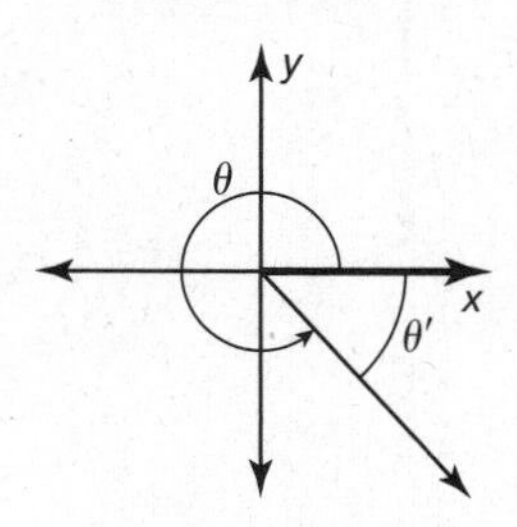

Degrees: $\theta' = 360° - \theta$

Radians: $\theta' = 2\pi - \theta$

Notes:

Evaluating Trigonometric Functions

Use these steps to evaluate a trigonometric function for any angle θ:

Step 1 Find the reference angle θ'.

Step 2 Evaluate the trigonometric function for θ'.

Step 3 Determine the sign of the trigonometric function value from the quadrant in which θ lies.

Signs of Function Values

Quadrant II	Quadrant I
$\sin\theta$, $\csc\theta$: +	$\sin\theta$, $\csc\theta$: +
$\cos\theta$, $\sec\theta$: −	$\cos\theta$, $\sec\theta$: +
$\tan\theta$, $\cot\theta$: −	$\tan\theta$, $\cot\theta$: +
Quadrant III	**Quadrant IV**
$\sin\theta$, $\csc\theta$: −	$\sin\theta$, $\csc\theta$: −
$\cos\theta$, $\sec\theta$: −	$\cos\theta$, $\sec\theta$: +
$\tan\theta$, $\cot\theta$: +	$\tan\theta$, $\cot\theta$: −

Notes:

Name ______________________________ Date __________

Extra Practice

In Exercises 1 and 2, evaluate the six trigonometric functions of θ.

1. **2.**

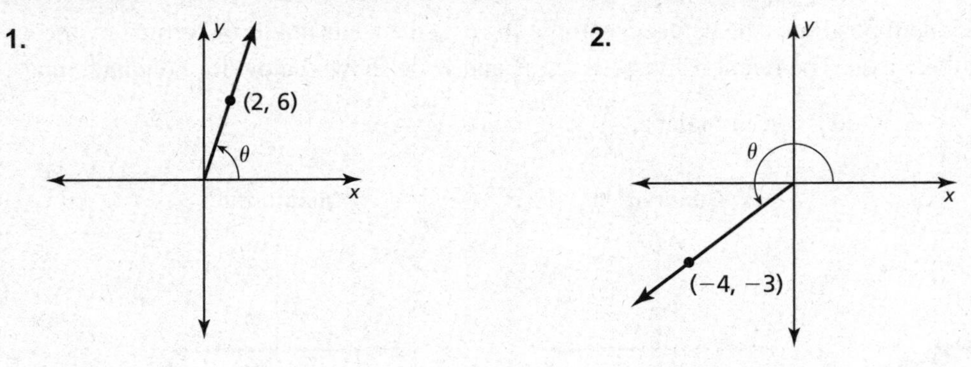

In Exercises 3 and 4, use the unit circle to evaluate the six trigonometric functions of θ.

3. $\theta = -90^\circ$

4. $\theta = 4\pi$

In Exercises 5 and 6, sketch the angle. Then find its reference angle.

5. -310°

6. $\dfrac{27\pi}{10}$

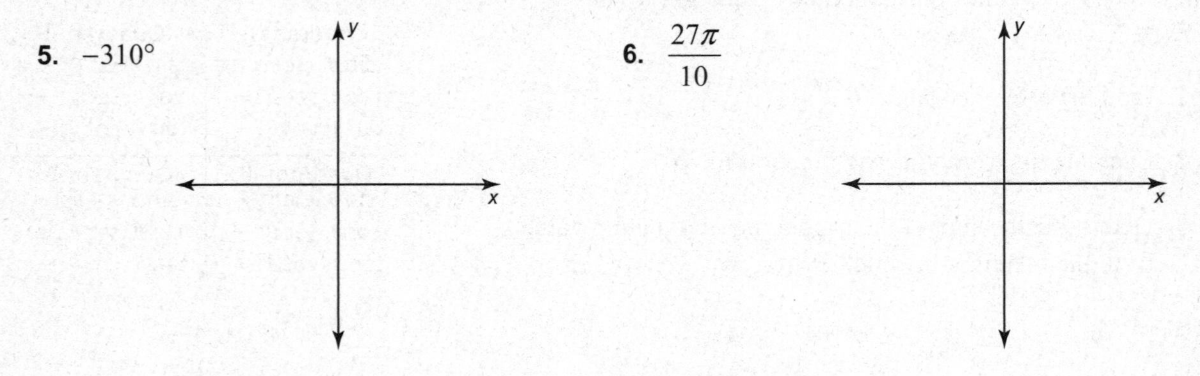

7. Evaluate the function csc 150° without using a calculator.

Name______________________________ Date__________

9.4 Graphing Sine and Cosine Functions

For use with Exploration 9.4

Essential Question What are the characteristics of the graphs of the sine and cosine functions?

EXPLORATION: Graphing the Sine Function

Go to *BigIdeasMath.com* for an interactive tool to investigate this exploration.

Work with a partner.

a. Complete the table for $y = \sin x$, where x is an angle measure in radians.

x	-2π	$-\frac{7\pi}{4}$	$-\frac{3\pi}{2}$	$-\frac{5\pi}{4}$	$-\pi$	$-\frac{3\pi}{4}$	$-\frac{\pi}{2}$	$-\frac{\pi}{4}$	0
$y = \sin x$									
x	$\frac{\pi}{4}$	$\frac{\pi}{2}$	$\frac{3\pi}{4}$	π	$\frac{5\pi}{4}$	$\frac{3\pi}{2}$	$\frac{7\pi}{4}$	2π	$\frac{9\pi}{4}$
$y = \sin x$									

b. Plot the points (x, y) from part (a). Draw a smooth curve through the points to sketch the graph of $y = \sin x$.

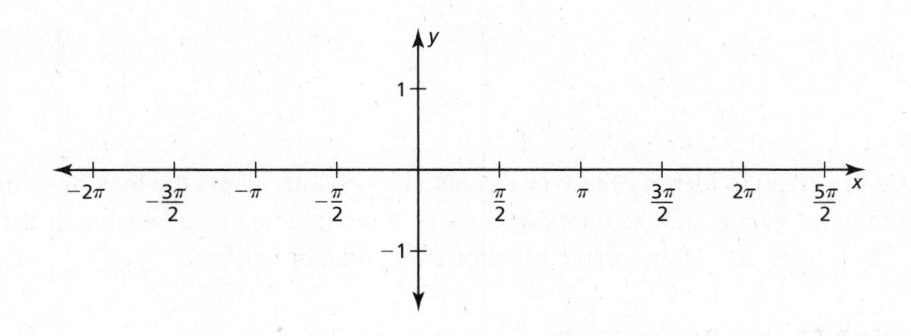

c. Use the graph to identify the x-intercepts, the x-values where the local maximums and minimums occur, and the intervals for which the function is increasing or decreasing over $-2\pi \leq x \leq 2\pi$. Is the sine function *even*, *odd*, or *neither*?

2 EXPLORATION: Graphing the Cosine Function

Go to *BigIdeasMath.com* for an interactive tool to investigate this exploration.

Work with a partner.

a. Complete the table for $y = \cos x$ using the same values of x as those used in Exploration 1.

x	-2π	$-\frac{7\pi}{4}$	$-\frac{3\pi}{2}$	$-\frac{5\pi}{4}$	$-\pi$	$-\frac{3\pi}{4}$	$-\frac{\pi}{2}$	$-\frac{\pi}{4}$	0
$y = \cos x$									
x	$\frac{\pi}{4}$	$\frac{\pi}{2}$	$\frac{3\pi}{4}$	π	$\frac{5\pi}{4}$	$\frac{3\pi}{2}$	$\frac{7\pi}{4}$	2π	$\frac{9\pi}{4}$
$y = \cos x$									

b. Plot the points (x, y) from part (a) and sketch the graph of $y = \cos x$

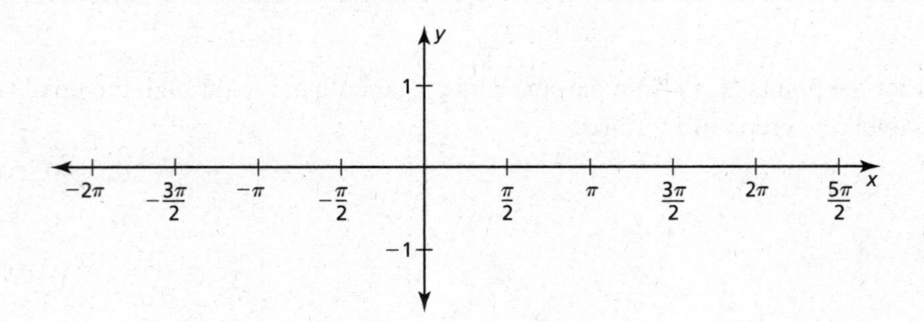

c. Use the graph to identify the x-intercepts, the x-values where the local maximums and minimums occur, and the intervals for which the function is increasing or decreasing over $-2\pi \leq x \leq 2\pi$. Is the cosine function *even*, *odd*, or *neither*?

Communicate Your Answer

3. What are the characteristics of the graphs of the sine and cosine functions?

4. Describe the end behavior of the graph of $y = \sin x$.

Name____ Date__________

9.4 Notetaking with Vocabulary
For use after Lesson 9.4

In your own words, write the meaning of each vocabulary term.

amplitude

periodic function

cycle

period

phase shift

midline

Core Concepts

Characteristics of $y = \sin x$ and $y = \cos x$

- The domain of each function is all real numbers.
- The range of each function is $-1 \le y \le 1$. So, the minimum value of each function is -1 and the maximum value is 1.
- The **amplitude** of the graph of each function is one-half of the difference of the maximum value and the minimum value, or $\frac{1}{2}[1-(-1)] = 1$.
- Each function is **periodic**, which means that its graph has a repeating pattern. The shortest repeating portion of the graph is called a **cycle**. The horizontal length of each cycle is called the **period**. The graph of each function has a period of 2π.
- The x-intercepts for $y = \sin x$ occur when $x = 0, \pm\pi, \pm2\pi, \pm3\pi, \ldots$.
- The x-intercepts for $y = \cos x$ occur when $x = \pm\frac{\pi}{2}, \pm\frac{3\pi}{2}, \pm\frac{5\pi}{2}, \pm\frac{7\pi}{2}, \ldots$.

Name ____ Date __________

9.4 Notetaking with Vocabulary (continued)

Amplitude and Period

The amplitude and period of the graphs of $y = a \sin bx$ and $y = a \cos bx$, where a and b are nonzero real numbers, are as follows:

$$\text{Amplitude} = |a| \qquad\qquad \text{Period} = \frac{2\pi}{|b|}$$

Notes:

Graphing $y = a \sin b(x - h) + k$ and $y = a \cos b(x - h) + k$

To graph $y = a \sin b(x - h) + k$ or $y = a \cos b(x - h) + k$ where $a > 0$ and $b > 0$, follow these steps:

Step 1 Identify the amplitude a, the period $\frac{2\pi}{b}$, the horizontal shift h, and the vertical shift k of the graph.

Step 2 Draw the horizontal line $y = k$, called the **midline** of the graph.

Step 3 Find the five key points by translating the key points of $y = a \sin bx$ or $y = a \cos bx$ horizontally h units and vertically k units.

Step 4 Draw the graph through the five translated key points.

Notes:

Name________________________________ Date__________

9.4 Notetaking with Vocabulary (continued)

Extra Practice

In Exercises 1–4, identify the amplitude and period of the function. Then graph the function and describe the graph of g as a transformation of the graph of its parent function.

1. $g(x) = \sin 2x$

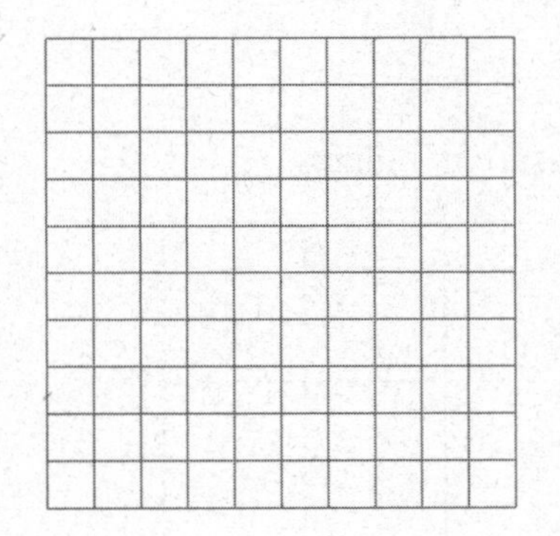

2. $g(x) = \frac{1}{3}\cos 2x$

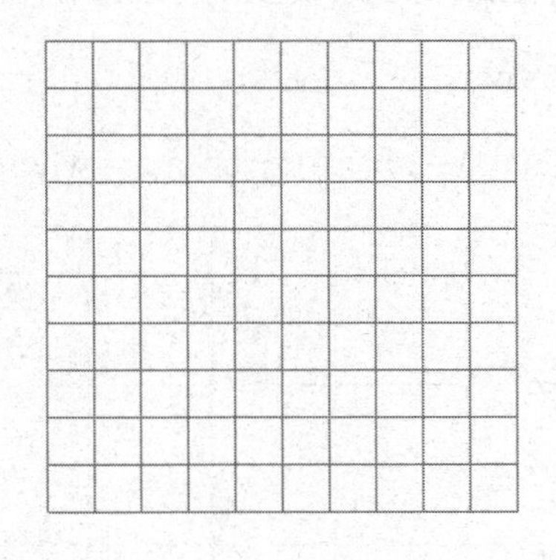

3. $g(x) = 4\sin 2\pi x$

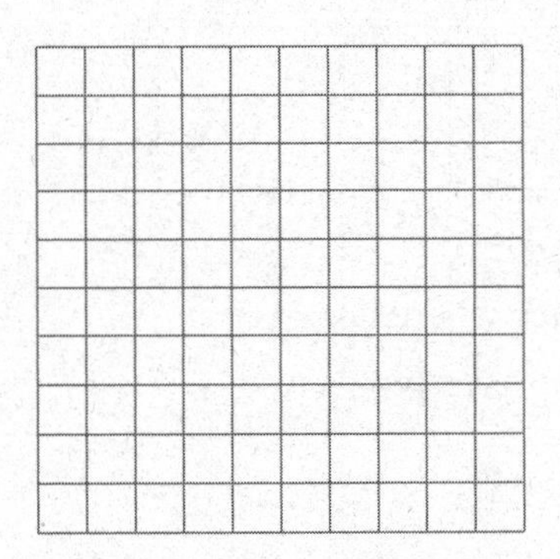

4. 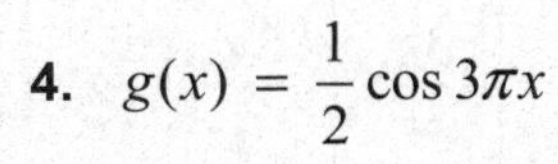 $g(x) = \frac{1}{2}\cos 3\pi x$

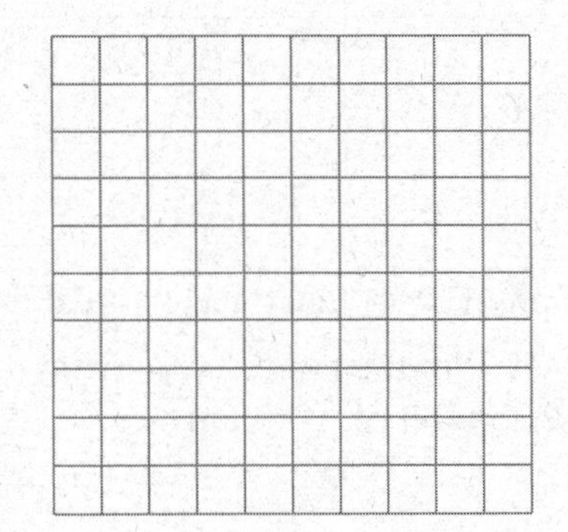

In Exercises 5 and 6, graph the function.

5. $g(x) = \sin \frac{1}{2}(x - \pi) + 1$

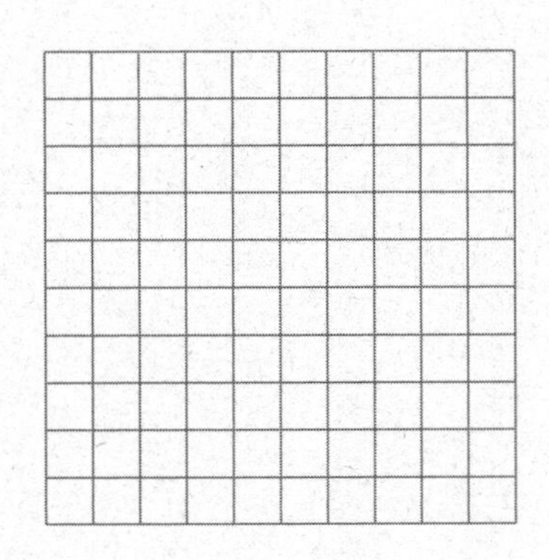

6. $g(x) = \cos\left(x + \frac{\pi}{2}\right) - 3$

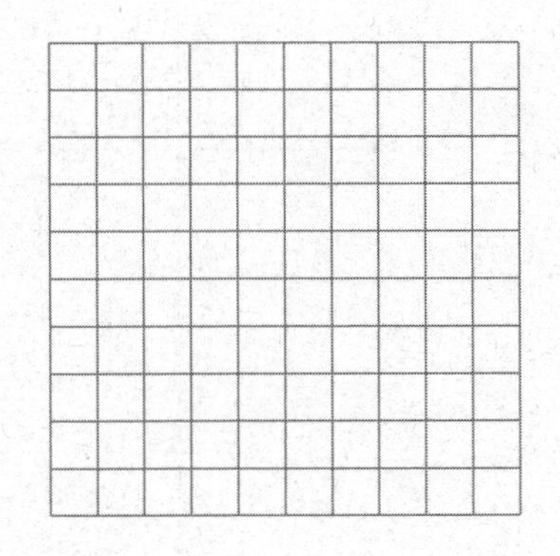

Name ______________________________ Date __________

9.5 Graphing Other Trigonometric Functions
For use with Exploration 9.5

Essential Question What are the characteristics of the graph of the tangent function?

1 EXPLORATION: Graphing the Tangent Function

Go to *BigIdeasMath.com* for an interactive tool to investigate this exploration.

Work with a partner.

a. Complete the table for $y = \tan x$, where x is an angle measure in radians.

x	$-\frac{\pi}{2}$	$-\frac{\pi}{3}$	$-\frac{\pi}{4}$	$-\frac{\pi}{6}$	0	$\frac{\pi}{6}$	$\frac{\pi}{4}$	$\frac{\pi}{3}$	$\frac{\pi}{2}$
$y = \tan x$									
x	$\frac{2\pi}{3}$	$\frac{3\pi}{4}$	$\frac{5\pi}{6}$	π	$\frac{7\pi}{6}$	$\frac{5\pi}{4}$	$\frac{4\pi}{3}$	$\frac{3\pi}{2}$	$\frac{5\pi}{3}$
$y = \tan x$									

b. The graph of $y = \tan x$ has vertical asymptotes at x-values where $\tan x$ is undefined. Plot the points (x, y) from part (a). Then use the asymptotes to sketch the graph of $y = \tan x$.

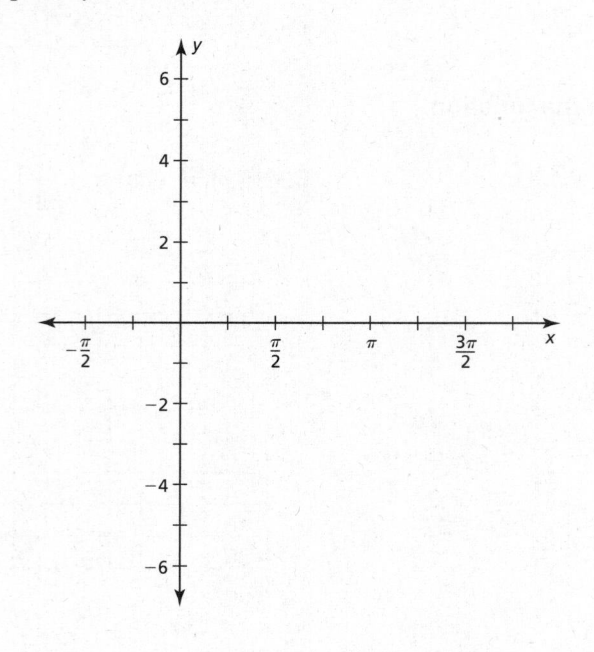

9.5 Graphing Other Trigonometric Functions (continued)

1 EXPLORATION: Graphing the Tangent Function (continued)

c. For the graph of $y = \tan x$, identify the asymptotes, the x-intercepts, and the intervals for which the function is increasing or decreasing over $-\frac{\pi}{2} \leq x \leq \frac{3\pi}{2}$. Is the tangent function *even*, *odd*, or *neither*?

Communicate Your Answer

2. What are the characteristics of the graph of the tangent function?

3. Describe the asymptotes of the graph of $y = \cot x$ on the interval $-\frac{\pi}{2} < x < \frac{3\pi}{2}$.

Name ____________________ Date ________

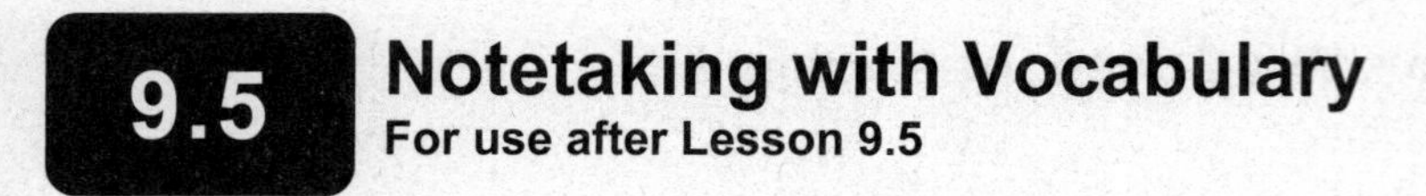

Notetaking with Vocabulary
For use after Lesson 9.5

In your own words, write the meaning of each vocabulary term.

asymptote

period

amplitude

x-intercept

transformations

Core Concepts

Characteristics of $y = \tan x$ and $y = \cot x$

The functions $y = \tan x$ and $y = \cot x$ have the following characteristics.

- The domain of $y = \tan x$ is all real numbers except odd multiples of $\frac{\pi}{2}$. At these x-values, the graph has vertical asymptotes.
- The domain of $y = \cot x$ is all real numbers except multiples of π. At these x-values, the graph has vertical asymptotes.
- The range of each function is all real numbers. So, the functions do not have maximum or minimum values, and the graphs do not have an amplitude.
- The period of each graph is π.
- The x-intercepts for $y = \tan x$ occur when $x = 0, \pm\pi, \pm 2\pi, \pm 3\pi, \ldots$.
- The x-intercepts for $y = \cot x$ occur when $x = \pm\frac{\pi}{2}, \pm\frac{3\pi}{2}, \pm\frac{5\pi}{2}, \pm\frac{7\pi}{2}, \ldots$.

Notes:

Period and Vertical Asymptotes of $y = a \tan bx$ and $y = a \cot bx$

The period and vertical asymptotes of the graphs of $y = a \tan bx$ and $y = a \cot bx$, where a and b are nonzero real numbers, are as follows.

- The period of the graph of each function is $\frac{\pi}{|b|}$.
- The vertical asymptotes for $y = a \tan bx$ are at odd multiples of $\frac{\pi}{2|b|}$.
- The vertical asymptotes for $y = a \cot bx$ are at multiples of $\frac{\pi}{|b|}$.

Notes:

Characteristics of $y = \sec x$ and $y = \csc x$

The functions $y = \sec x$ and $y = \csc x$ have the following characteristics.

- The domain of $y = \sec x$ is all real numbers except odd multiples of $\frac{\pi}{2}$. At these x-values, the graph has vertical asymptotes.
- The domain of $y = \csc x$ is all real numbers except multiples of π. At these x-values, the graph has vertical asymptotes.
- The range of each function is $y \leq -1$ and $y \geq 1$. So, the graphs do not have an amplitude.
- The period of each graph is 2π.

Notes:

Name ____ Date __________

Extra Practice

In Exercises 1–6, graph one period of the function. Describe the graph of *g* as a transformation of the graph of its parent function.

1. $g(x) = \tan 2x$

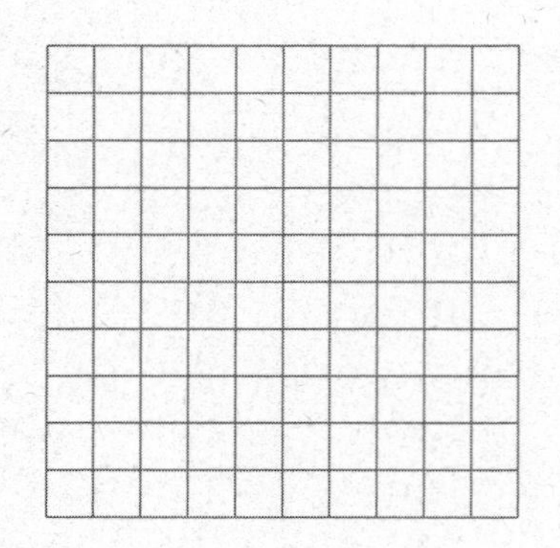

2. $g(x) = 2 \cot \frac{1}{2}x$

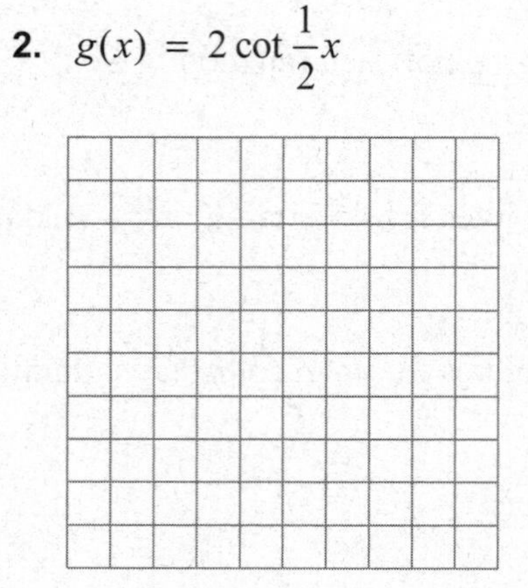

3. $g(x) = \frac{1}{4} \tan \frac{\pi}{4}x$

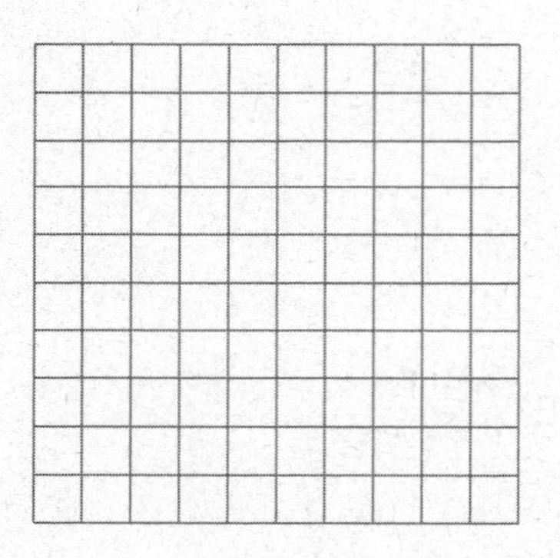

4. $g(x) = \frac{1}{2} \cot 3x$

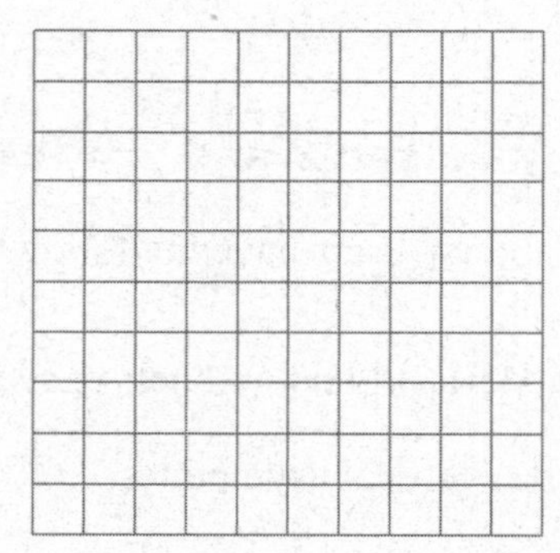

5. $g(x) = 2 \sec 2x$

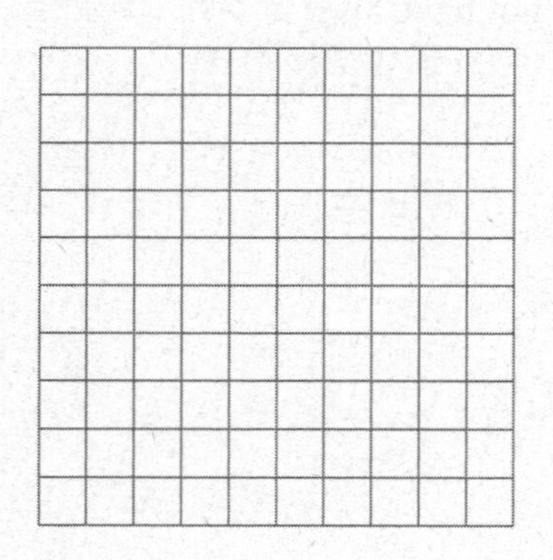

6. $g(x) = \csc 2\pi x$

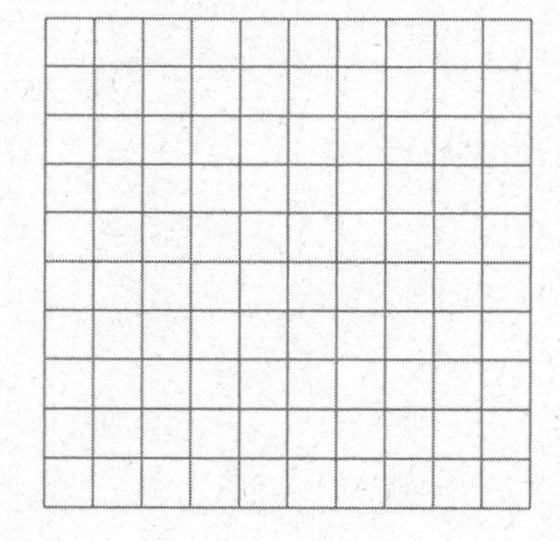

Name____ Date__________

9.6 Modeling with Trigonometric Functions

For use with Exploration 9.6

Essential Question What are the characteristics of the real-life problems that can be modeled by trigonometric functions?

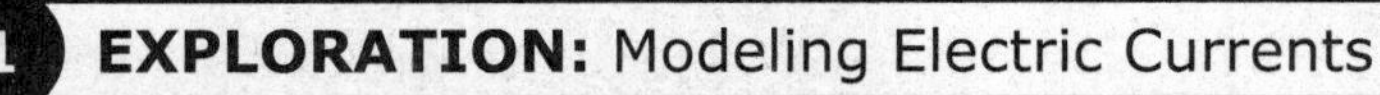

1 EXPLORATION: Modeling Electric Currents

Work with a partner. Find a sine function that models the electric current shown in each oscilloscope screen. State the amplitude and period of the graph.

a.

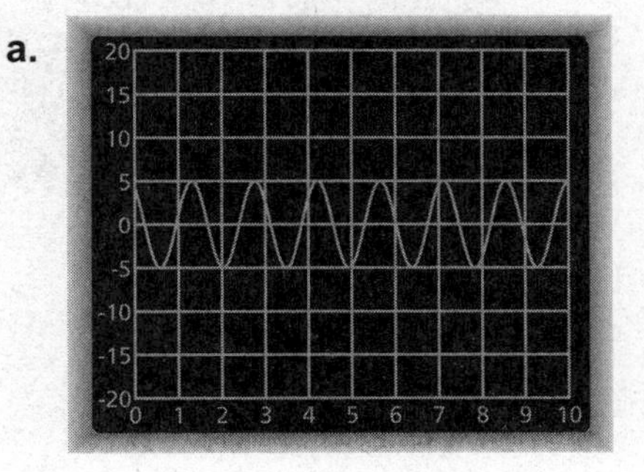

b.

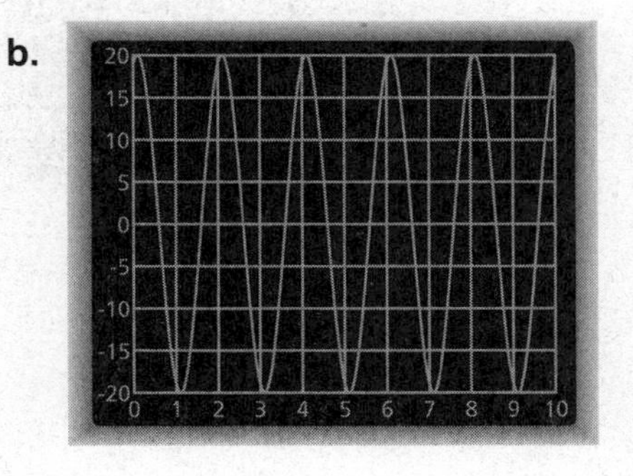

c.

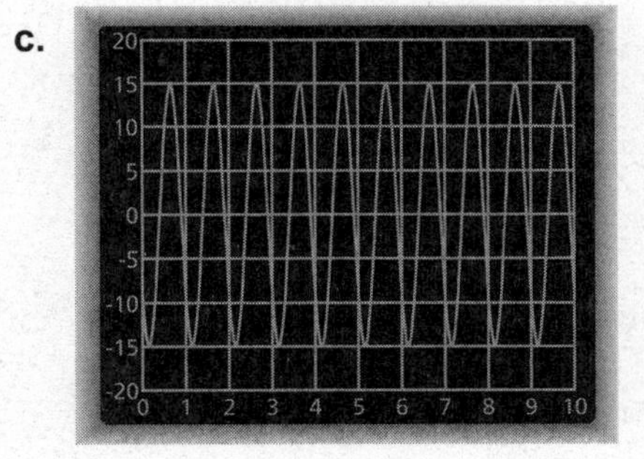

d.

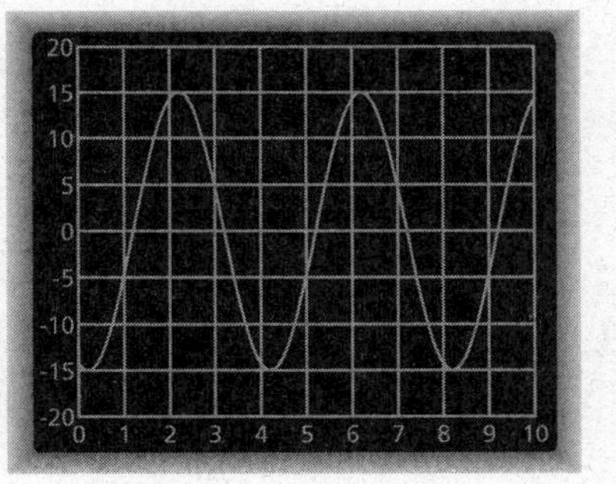

Name ____ Date __________

9.6 Modeling with Trigonometric Functions (continued)

1 EXPLORATION: Modeling Electric Currents (continued)

e. f.

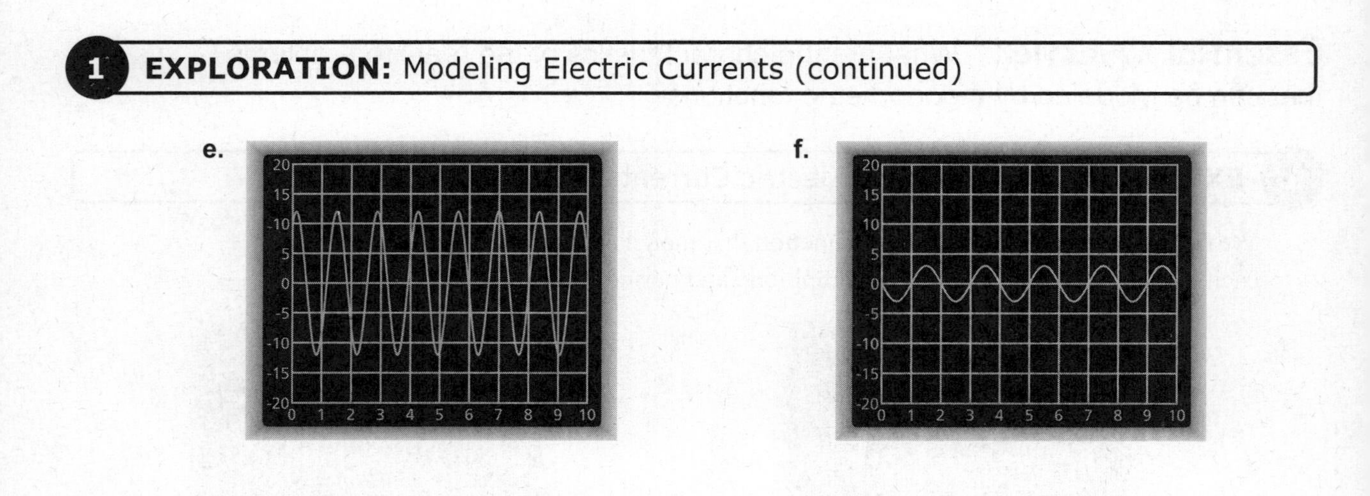

Communicate Your Answer

2. What are the characteristics of the real-life problems that can be modeled by trigonometric functions?

3. Use the Internet or some other reference to find examples of real-life situations that can be modeled by trigonometric functions.

Name____ Date__________

9.6 Notetaking with Vocabulary
For use after Lesson 9.6

In your own words, write the meaning of each vocabulary term.

frequency

sinusoid

Notes:

Extra Practice

1. An alternating current generator (AC generator) converts motion to electricity by generating sinusoidal voltage. Assuming that there is no vertical offset and phase shift, the voltage oscillates between −170 volts and +170 volts with a frequency of 60 hertz. Write and graph a sine model that gives the voltage V as a function of the time t (in seconds).

In Exercises 2–5, write a function for the sinusoid.

2.

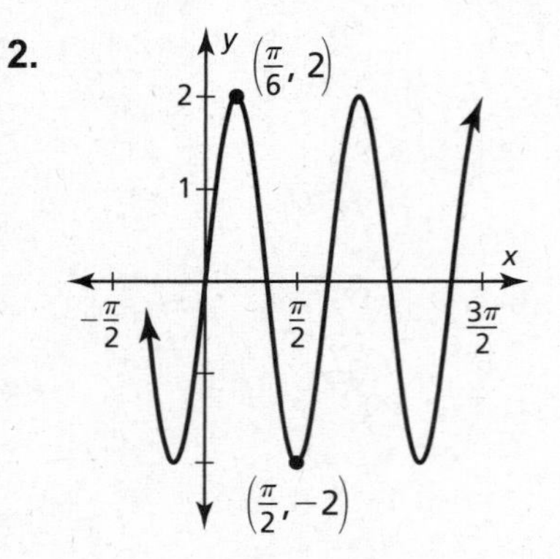

3.

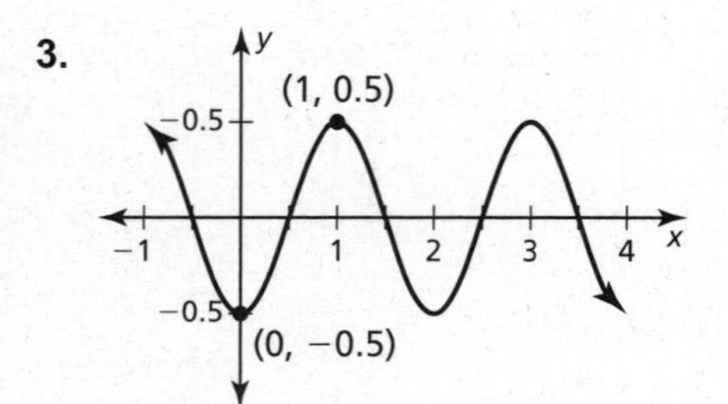

4.

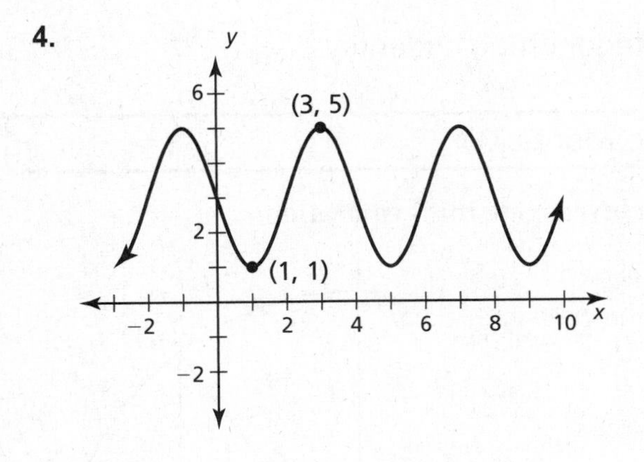

5.

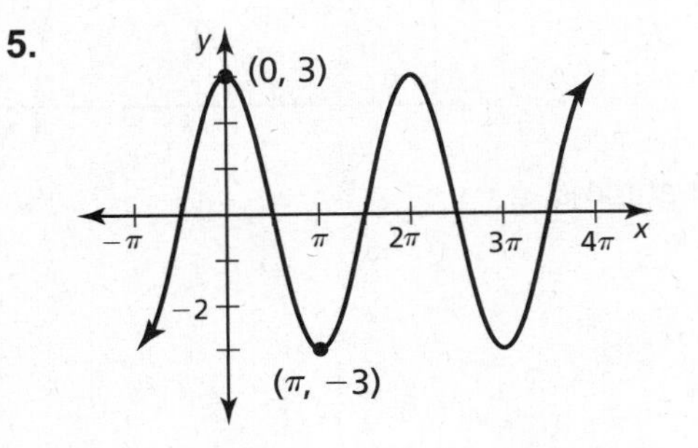

6. The pedal of a bicycle wheel is 7 inches long. The lowest point of the pedal is 4 inches above the ground. A cyclist pedals 3 revolutions per second. Write a model for the height h (in inches) of the pedal as a function of the time t (in seconds) given that the pedal is at its lowest point when $t = 0$.

7. The London Eye, the tallest Ferris wheel in Europe, has a diameter of 120 meters and the whole structure is 135 meters tall. The Ferris wheel completes one revolution in about 30 minutes. Write a model for the height h (in meters) of a passenger capsule as a function of the time t (in seconds) given that the capsule is at its highest point when $t = 0$.

Name ________________________________ Date __________

9.7 Using Trigonometric Identities
For use with Exploration 9.7

Essential Question How can you verify a trigonometric identity?

1 EXPLORATION: Writing a Trigonometric Identity

Go to *BigIdeasMath.com* for an interactive tool to investigate this exploration.

Work with a partner. In the figure, the point (x, y) is on a circle of radius c with center at the origin.

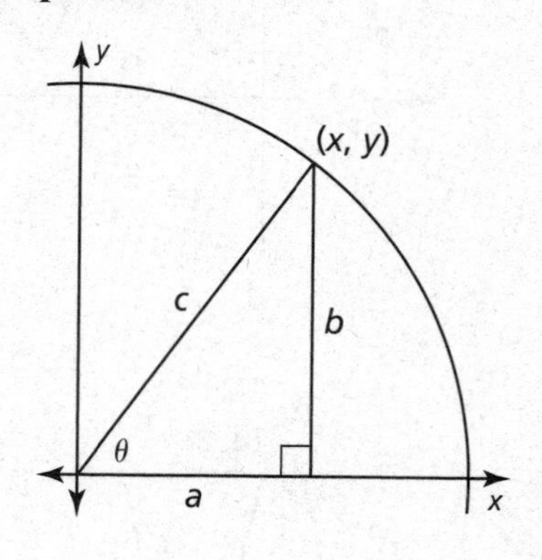

a. Write an equation that relates a, b, and c.

b. Write expressions for the sine and cosine ratios of angle θ.

c. Use the results from parts (a) and (b) to find the sum of $\sin^2\theta$ and $\cos^2\theta$. What do you observe?

d. Complete the table to verify that the identity you wrote in part (c) is valid for angles (of your choice) in each of the four quadrants.

	θ	$\sin^2\theta$	$\cos^2\theta$	$\sin^2\theta + \cos^2\theta$
QI				
QII				
QIII				
QIV				

2 EXPLORATION: Writing Other Trigonometric Identities

Work with a partner. The trigonometric identity you derived in Exploration 1 is called a Pythagorean identity. There are two other Pythagorean identities. To derive them, recall the four relationships:

$$\tan\theta = \frac{\sin\theta}{\cos\theta} \qquad \cot\theta = \frac{\cos\theta}{\sin\theta}$$

$$\sec\theta = \frac{1}{\cos\theta} \qquad \csc\theta = \frac{1}{\sin\theta}$$

a. Divide each side of the Pythagorean identity you derived in Exploration 1 by $\cos^2\theta$ and simplify. What do you observe?

b. Divide each side of the Pythagorean identity you derived in Exploration 1 by $\sin^2\theta$ and simplify. What do you observe?

Communicate Your Answer

3. How can you verify a trigonometric identity?

4. Is $\sin\theta = \cos\theta$ a trigonometric identity? Explain your reasoning.

5. Give some examples of trigonometric identities that are different than those in Explorations 1 and 2.

Name ______________________________ Date __________

9.7 Notetaking with Vocabulary

For use after Lesson 9.7

In your own words, write the meaning of each vocabulary term.

trigonometric identity

Core Concepts

Fundamental Trigonometric Identities

Reciprocal Identities

$\csc \theta = \dfrac{1}{\sin \theta}$ $\sec \theta = \dfrac{1}{\cos \theta}$ $\cot \theta = \dfrac{1}{\tan \theta}$

Tangent and Cotangent Identities

$\tan \theta = \dfrac{\sin \theta}{\cos \theta}$ $\cot \theta = \dfrac{\cos \theta}{\sin \theta}$

Pythagorean Identities

$\sin^2 \theta + \cos^2 \theta = 1$ $1 + \tan^2 \theta = \sec^2 \theta$ $1 + \cot^2 \theta = \csc^2 \theta$

Cofunction Identities

$\sin\left(\dfrac{\pi}{2} - \theta\right) = \cos \theta$ $\cos\left(\dfrac{\pi}{2} - \theta\right) = \sin \theta$ $\tan\left(\dfrac{\pi}{2} - \theta\right) = \cot \theta$

Negative Angle Identities

$\sin(-\theta) = -\sin \theta$ $\cos(-\theta) = \cos \theta$ $\tan(-\theta) = -\tan \theta$

Notes:

9.7 Notetaking with Vocabulary (continued)

Extra Practice

In Exercises 1–4, find the values of the other five trigonometric functions of θ.

1. $\sin\theta = \frac{1}{5}, \frac{\pi}{2} < \theta < \pi$

2. $\cos\theta = -\frac{4}{5}, \pi < \theta < \frac{3\pi}{2}$

3. $\cot\theta = \frac{4}{7}, 0 < \theta < \frac{\pi}{2}$

4. $\sec\theta = \frac{\sqrt{13}}{3}, \frac{3\pi}{2} < \theta < 2\pi$

In Exercises 5–8, simplify the expression.

5. $-\frac{\tan\theta}{\sec\theta}$

6. $\cos\left(\frac{\pi}{2} - \theta\right)(1 - \cos^2\theta)$

Name ______________________________ Date __________

9.7 Notetaking with Vocabulary (continued)

7. $\dfrac{2\sec^2 x - 2\tan^2 x}{\tan(-x)\cos(-x)}$

8. $\dfrac{-\sin\left(\dfrac{\pi}{2} - \theta\right)}{\sec(-\theta)} - \sin^2\theta$

In Exercises 9 and 10, verify the identity.

9. $\dfrac{1 - \cos^2\theta}{\sec^2\theta} - \sin^2\theta = -\sin^4\theta$

10. $\csc x + \cot x = \dfrac{\sin x}{1 - \cos x}$

Name____ Date__________

9.8 Using Sum and Difference Formulas

For use with Exploration 9.8

Essential Question How can you evaluate trigonometric functions of the sum or difference of two angles?

1 EXPLORATION: Deriving a Difference Formula

Work with a partner.

a. Explain why the two triangles shown are congruent.

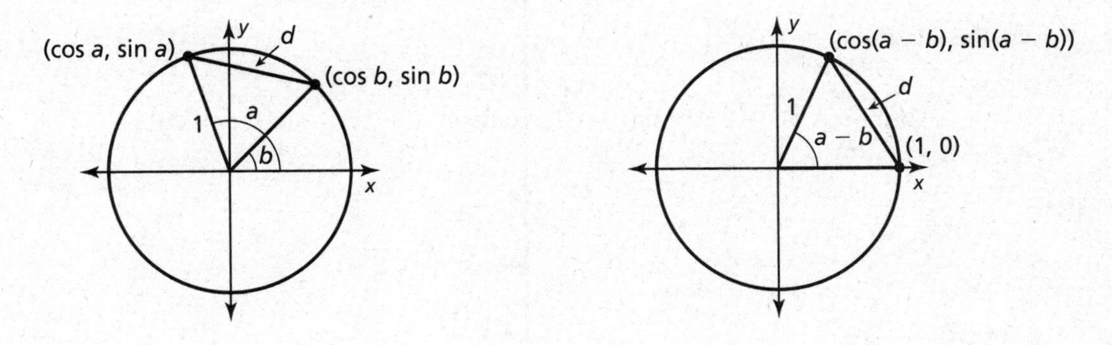

b. Use the Distance Formula to write an expression for d in the first unit circle.

c. Use the Distance Formula to write an expression for d in the second unit circle.

d. Write an equation that relates the expressions in parts (b) and (c). Then simplify this equation to obtain a formula for $\cos(a - b)$.

2 EXPLORATION: Deriving a Sum Formula

Work with a partner. Use the difference formula you derived in Exploration 1 to write a formula for $\cos(a + b)$ in terms of sine and cosine of a and b. *Hint*: Use the fact that

$$\cos(a + b) = \cos[a - (-b)].$$

3 EXPLORATION: Deriving Difference and Sum Formulas

Work with a partner. Use the formulas you derived in Explorations 1 and 2 to write formulas for $\sin(a - b)$ and $\sin(a + b)$ in terms of sine and cosine of a and b. *Hint*: Use the cofunction identities

$$\sin\left(\frac{\pi}{2} - a\right) = \cos a \text{ and } \cos\left(\frac{\pi}{2} - a\right) = \sin a$$

and the fact that

$$\cos\left[\left(\frac{\pi}{2} - a\right) + b\right] = \sin(a - b) \text{ and } \sin(a + b) = \sin[a - (-b)].$$

Communicate Your Answer

4. How can you evaluate trigonometric functions of the sum or difference of two angles?

5. a. Find the exact values of sin 75° and cos 75° using sum formulas. Explain your reasoning.

b. Find the exact values of sin 75° and cos 75° using difference formulas. Compare your answers to those in part (a).

Name____ Date__________

9.8 Notetaking with Vocabulary
For use after Lesson 9.8

In your own words, write the meaning of each vocabulary term.

ratio

Core Concepts

Sum and Difference Formulas

Sum Formulas

$\sin(a + b) = \sin a \cos b + \cos a \sin b$

$\cos(a + b) = \cos a \cos b - \sin a \sin b$

$\tan(a + b) = \dfrac{\tan a + \tan b}{1 - \tan a \tan b}$

Difference Formulas

$\sin(a - b) = \sin a \cos b - \cos a \sin b$

$\cos(a - b) = \cos a \cos b + \sin a \sin b$

$\tan(a - b) = \dfrac{\tan a - \tan b}{1 + \tan a \tan b}$

Notes:

Name ____ Date __________

Extra Practice

In Exercises 1–4, find the exact value of the expression.

1. $\sin(-75^\circ)$

2. $\tan 120^\circ$

3. $\cos\left(-\dfrac{7\pi}{12}\right)$

4. $\cos \dfrac{35\pi}{12}$

In Exercises 5–8, evaluate the expression given that $\sin a = -\frac{4}{5}$ with $\pi < a < \frac{3\pi}{2}$ and $\cos b = \frac{5}{13}$ with $0 < b < \frac{\pi}{2}$.

5. $\cos(a - b)$

6. $\sin(a + b)$

Name______________________________ Date__________

9.8 Notetaking with Vocabulary (continued)

7. $\tan(a + b)$

8. $\tan(a - b)$

In Exercises 9–12, simplify the expression.

9. $\sin\left(x + \frac{\pi}{2}\right)$

10. $\tan(x - 2\pi)$

11. $\cos(x - 2\pi)$

12. $\cos\left(x + \frac{5\pi}{2}\right)$

In Exercises 13 and 14, solve the equation for $0 \le x \le 2\pi$.

13. $\sin\left(x + \frac{3\pi}{2}\right) = 1$

14. $\sin\left(x - \frac{\pi}{2}\right) + \cos\left(x - \frac{\pi}{2}\right) = 0$

Name ______________________________ Date __________

Chapter 10 Maintaining Mathematical Proficiency

Write and solve a proportion to answer the question.

1. What percent of 260 is 65?

2. What number is 32% of 75?

3. 15.01 is what percent of 19?

Display the data in a histogram.

4.

	Number of Strike-outs in One Game		
Strike-outs	0–3	4–7	8–11
Frequency	34	20	8

5.

	Number of Days of Exercise in One Week			
Days of Exercise	0–1	2–3	4–5	6–7
Frequency	4	26	22	6

Name____ Date__________

10.1 Sample Spaces and Probability

For use with Exploration 10.1

Essential Question How can you list the possible outcomes in the sample space of an experiment?

The **sample space** of an experiment is the set of all possible outcomes for that experiment.

1 EXPLORATION: Finding the Sample Space of an Experiment

Go to *BigIdeasMath.com* for an interactive tool to investigate this exploration.

Work with a partner. In an experiment, three coins are flipped. List the possible outcomes in the sample space of the experiment.

2 EXPLORATION: Finding the Sample Space of an Experiment

Go to Big*IdeasMath.com* for an interactive tool to investigate this exploration.

Work with a partner. List the possible outcomes in the sample space of the experiment.

a. One six-sided die is rolled.

b. Two six-sided dice are rolled.

3 EXPLORATION: Finding the Sample Space of an Experiment

Go to *BigIdeasMath.com* for an interactive tool to investigate this exploration.

Work with a partner. In an experiment, a spinner is spun.

a. How many ways can you spin a 1? 2? 3? 4? 5?

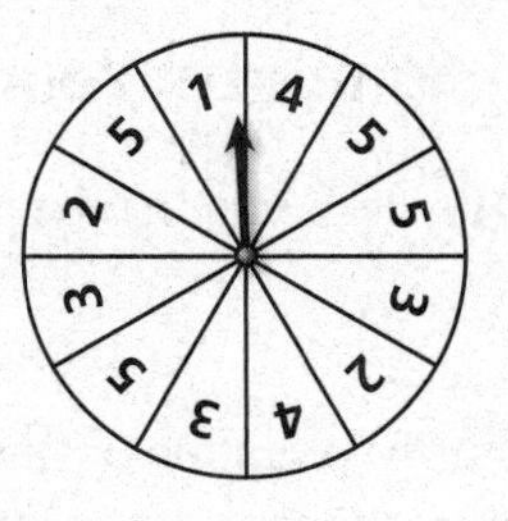

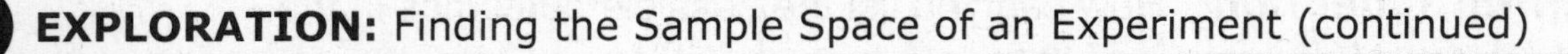

3 EXPLORATION: Finding the Sample Space of an Experiment (continued)

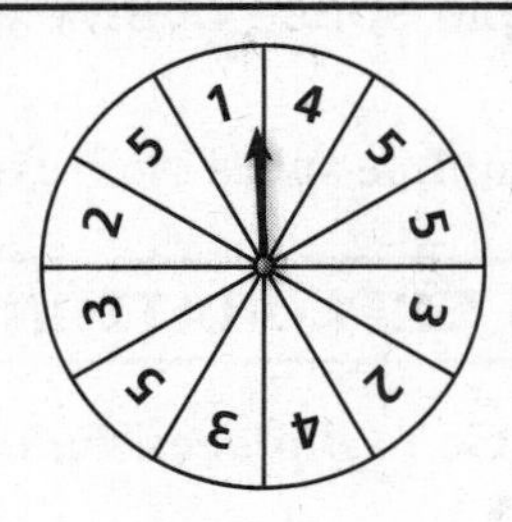

b. List the sample space.

c. What is the total number of outcomes?

4 EXPLORATION: Finding the Sample Space of an Experiment

Go to *BigIdeasMath.com* for an interactive tool to investigate this exploration.

Work with a partner. In an experiment, a bag contains 2 blue marbles and 5 red marbles. Two marbles are drawn from the bag.

a. How many ways can you choose two blue? a red then blue? a blue then red? two red?

b. List the sample space.

c. What is the total number of outcomes?

Communicate Your Answer

5. How can you list the possible outcomes in the sample space of an experiment?

6. For Exploration 3, find the ratio of the number of each possible outcome to the total number of outcomes. Then find the sum of these ratios. Repeat for Exploration 4. What do you observe?

Name____ Date__________

10.1 Notetaking with Vocabulary

For use after Lesson 10.1

In your own words, write the meaning of each vocabulary term.

probability experiment

outcome

event

sample space

probability of an event

theoretical probability

geometric probability

experimental probability

Core Concepts

Probability of the Complement of an Event

The probability of the complement of event A is

$$P(\overline{A}) = 1 - P(A).$$

Notes:

Name ______________________ Date __________

10.1 Notetaking with Vocabulary (continued)

Extra Practice

In Exercises 1 and 2, find the number of possible outcomes in the sample space. Then list the possible outcomes.

1. A stack of cards contains the thirteen clubs from a standard deck of cards. You pick a card from the stack and flip two coins.

2. You spin a spinner with the numbers 1–5 on it and roll a die.

3. When two tiles with numbers between 1 and 10 are chosen from two different bags, there are 100 possible outcomes. Find the probability that (a) the sum of the two numbers is not 10 and (b) the product of the numbers is greater than 10.

4. At a school dance, the parents sell pizza slices. The table shows the number of pizza slices that are available. A student chooses a slice at random. What is the probability that the student chooses a thin crust slice with pepperoni?

	Pepperoni	Plain Cheese
Thin Crust	34	36
Thick Crust	8	12

5. Find the probability that the polynomial $x^2 - x - c$ can be factored if c is a randomly chosen integer from 1 to 12.

6. You throw a dart at the board shown. Your dart is equally likely to hit any point inside the square board.

a. What is the probability your dart lands in the smallest triangle?

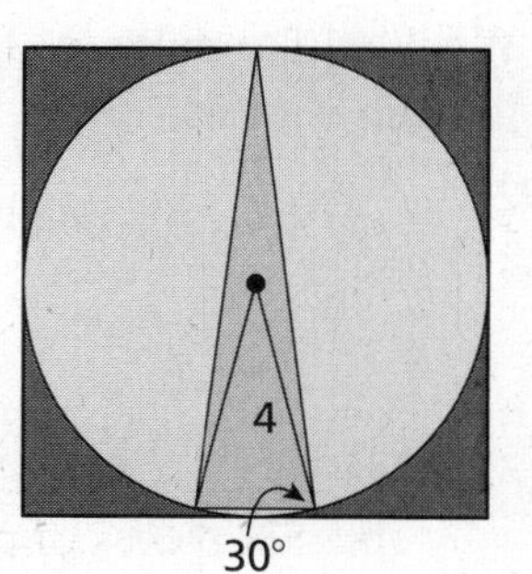

b. What is the probability your dart does not land anywhere in the circle?

7. The sections of a spinner are numbered 1 through 12. Each section of the spinner has the same area. You spin the spinner 180 times. The table shows the results. For which number is the experimental probability of stopping on the number the same as the theoretical probability?

Spinner Results											
1	**2**	**3**	**4**	**5**	**6**	**7**	**8**	**9**	**10**	**11**	**12**
13	21	22	20	11	8	14	9	15	12	18	17

Name ______________________________ Date __________

10.2 Independent and Dependent Events

For use with Exploration 10.2

Essential Question How can you determine whether two events are independent or dependent?

Two events are **independent events** when the occurrence of one event does not affect the occurrence of the other event. Two events are **dependent events** when the occurrence of one event *does* affect the occurrence of the other event.

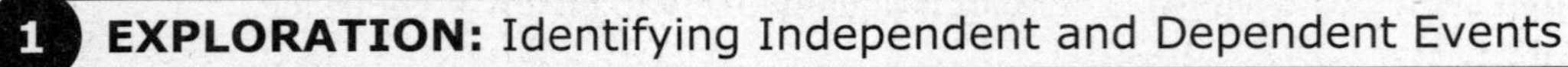

1 EXPLORATION: Identifying Independent and Dependent Events

Work with a partner. Determine whether the events are independent or dependent. Explain your reasoning.

a. Two six-sided dice are rolled.

b. Six pieces of paper, numbered 1 through 6, are in a bag. Two pieces of paper are selected one at a time without replacement.

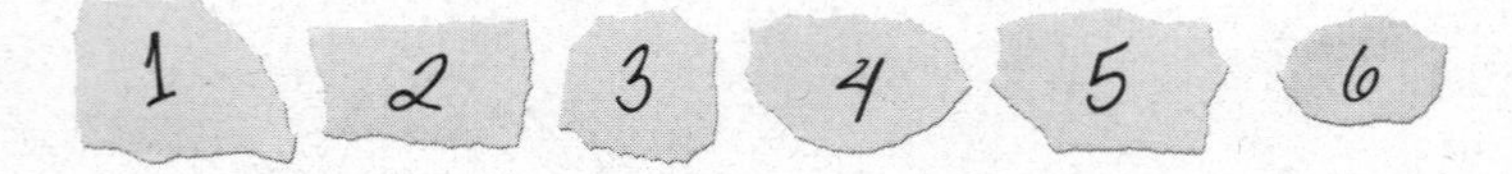

2 EXPLORATION: Finding Experimental Probabilities

Go to *BigIdeasMath.com* for an interactive tool to investigate this exploration.

Work with a partner.

a. In Exploration 1(a), experimentally estimate the probability that the sum of the two numbers rolled is 7. Describe your experiment.

b. In Exploration 1(b), experimentally estimate the probability that the sum of the two numbers selected is 7. Describe your experiment.

3 EXPLORATION: Finding Theoretical Probabilities

Work with a partner.

a. In Exploration 1(a), find the theoretical probability that the sum of the two numbers rolled is 7. Then compare your answer with the experimental probability you found in Exploration 2(a).

b. In Exploration 1(b), find the theoretical probability that the sum of the two numbers selected is 7. Then compare your answer with the experimental probability you found in Exploration 2(b).

c. Compare the probabilities you obtained in parts (a) and (b).

Communicate Your Answer

4. How can you determine whether two events are independent or dependent?

5. Determine whether the events are independent or dependent. Explain your reasoning.

a. You roll a 4 on a six-sided die and spin red on a spinner.

b. Your teacher chooses a student to lead a group, chooses another student to lead a second group, and chooses a third student to lead a third group.

Name ______________________________ Date __________

10.2 Notetaking with Vocabulary

For use after Lesson 10.2

In your own words, write the meaning of each vocabulary term.

independent events

dependent events

conditional probability

Core Concepts

Probability of Independent Events

Words Two events A and B are independent events if and only if the probability that both events occur is the product of the probabilities of the events.

Symbols $P(A \text{ and } B) = P(A) \bullet P(B)$

Notes:

Probability of Dependent Events

Words If two events A and B are dependent events, then the probability that both events occur is the product of the probability of the first event and the conditional probability of the second event given the first event.

Symbols $P(A \text{ and } B) = P(A) \bullet P(B|A)$

Example Using the information in Example 2:

$$P(\text{girl first and girl second}) = P(\text{girl first}) \bullet P(\text{girl second}|\text{girl first})$$

$$= \frac{9}{12} \bullet \frac{6}{9} = \frac{1}{2}$$

Notes:

Extra Practice

In Exercises 1 and 2, determine whether the events are independent. Explain your reasoning.

1. You have three white golf balls and two yellow golf balls in a bag. You randomly select one golf ball to hit now and another golf ball to place in your pocket. Use a sample space to determine whether randomly selecting a white golf ball first and then a white golf ball second are independent events.

2. Your friend writes a phone number down on a piece of paper but the last three numbers get smudged after being in your pocket all day long. You decide to randomly choose numbers for each of the three digits. Use a sample space to determine whether guessing the first digit correctly and the second digit correctly are independent events.

Name ______________________________ Date __________

10.2 Notetaking with Vocabulary (continued)

3. You are trying to guess a three-letter password that uses only the letters A, E, I, O, U, and Y. Letters can be used more than once. Find the probability that you pick the correct password "YOU."

4. You are trying to guess a three-letter password that uses only the letters A, E, I, O, U, and Y. Letters *cannot* be used more than once. Find the probability that you pick the correct password "AIE."

5. The table shows the number of male and female college students who played collegiate basketball and collegiate soccer in the United States in a recent year.

	Collegiate Soccer	**Collegiate Basketball**
Male	37,240	31,863
Female	36,523	28,002

a. Find the probability that a randomly selected collegiate soccer player is female.

b. Find the probability that a randomly selected male student is a collegiate basketball player.

Name____ Date__________

10.3 Two-Way Tables and Probability

For use with Exploration 10.3

Essential Question How can you construct and interpret a two-way table?

1 EXPLORATION: Completing and Using a Two-Way Table

Work with a partner. A *two-way table* displays the same information as a Venn diagram. In a two-way table, one category is represented by the rows and the other category is represented by the columns.

The Venn diagram shows the results of a survey in which 80 students were asked whether they play a musical instrument and whether they speak a foreign language. Use the Venn diagram to complete the two-way table. Then use the two-way table to answer each question.

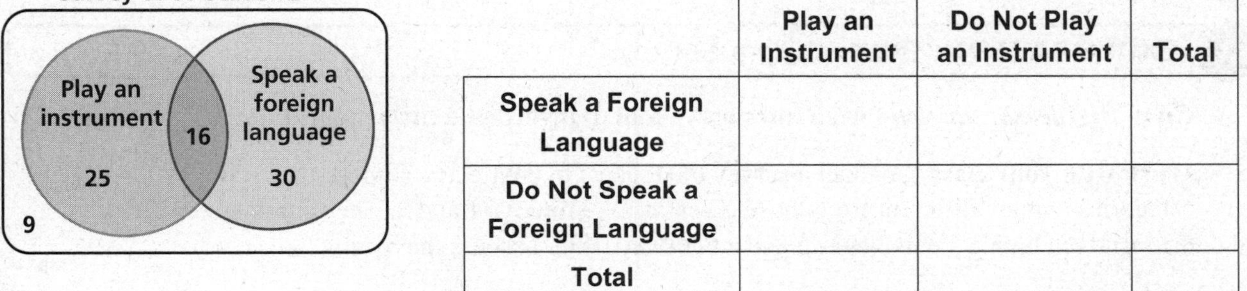

	Play an Instrument	Do Not Play an Instrument	Total
Speak a Foreign Language			
Do Not Speak a Foreign Language			
Total			

a. How many students play an instrument?

b. How many students speak a foreign language?

c. How many students play an instrument and speak a foreign language?

d. How many students do not play an instrument and do not speak a foreign language?

e. How many students play an instrument and do not speak a foreign language?

2 EXPLORATION: Two-Way Tables and Probability

Work with a partner. In Exploration 1, one student is selected at random from the 80 students who took the survey. Find the probability that the student

a. plays an instrument.

Name ______________________________ Date __________

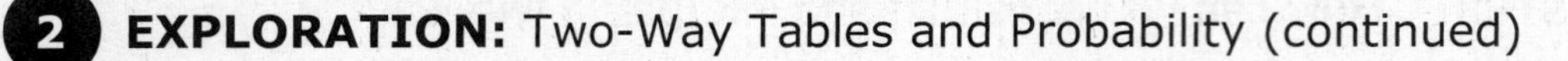

2 EXPLORATION: Two-Way Tables and Probability (continued)

b. speaks a foreign language.

c. plays an instrument and speaks a foreign language.

d. does not play an instrument and does not speak a foreign language.

e. plays an instrument and does not speak a foreign language.

3 EXPLORATION: Conducting a Survey

Go to *BigIdeasMath.com* for an interactive tool to investigate this exploration.

Work with your class. Conduct a survey of students in your class. Choose two categories that are different from those given in Explorations 1 and 2. Then summarize the results in both a Venn diagram and a two-way table. Discuss the results.

Communicate Your Answer

4. How can you construct and interpret a two-way table?

5. How can you use a two-way table to determine probabilities?

Name____ Date__________

10.3 Notetaking with Vocabulary
For use after Lesson 10.3

In your own words, write the meaning of each vocabulary term.

two-way table

joint frequency

marginal frequency

joint relative frequency

marginal relative frequency

conditional relative frequency

Core Concepts

Relative and Conditional Relative Frequencies

A **joint relative frequency** is the ratio of a frequency that is not in the total row or the total column to the total number of values or observations.

A **marginal relative frequency** is the sum of the joint relative frequencies in a row or a column.

A **conditional relative frequency** is the ratio of a joint relative frequency to the marginal relative frequency. You can find a conditional relative frequency using a row total or a column total of a two-way table.

Notes:

Name ____ Date __________

10.3 Notetaking with Vocabulary (continued)

Extra Practice

In Exercises 1 and 2, complete the two-way table.

1.

		Arrival		
		Tardy	On Time	Total
Method	Walk	22		
	City Bus			60
	Total		58	130

2.

		Response		
		Yes	No	Total
Age	Under 21		24	25
	Over 21	29		
	Total	30		75

3. A survey was taken of 100 families with one child and 86 families with two or more children to determine whether they were saving for college. Of those, 94 of the families with one child and 60 of the families with two or more children were saving for college. Organize these results in a two-way table. Then find and interpret the marginal frequencies.

10.3 Notetaking with Vocabulary (continued)

4. In a survey, 214 ninth graders played video games every day of the week and 22 ninth graders did not play video games every day of the week. Of those that played every day of the week, 36 had trouble sleeping at night. Of those that did not play every day of the week, 7 had trouble sleeping at night. Make a two-way table that shows the joint and marginal relative frequencies.

5. For financial reasons, a school district is debating about eliminating a Computer Programming class at the high school. The district surveyed parents, students, and teachers. The results, given as joint relative frequencies, are shown in the two-way table.

		Population		
		Parents	**Students**	**Teachers**
Response	**Yes**	0.58	0.08	0.10
	No	0.06	0.15	0.03

a. What is the probability that a randomly selected parent voted to eliminate the class?

b. What is the probability that a randomly selected student did not want to eliminate the class?

c. Determine whether voting to eliminate the class and being a teacher are independent events.

Name ______________________________ Date __________

10.4 Probability of Disjoint and Overlapping Events

For use with Exploration 10.4

Essential Question How can you find probabilities of disjoint and overlapping events?

Two events are **disjoint**, or **mutually exclusive**, when they have no outcomes in common. Two events are **overlapping** when they have one or more outcomes in common.

1 EXPLORATION: Disjoint Events and Overlapping Events

Go to *BigIdeasMath.com* for an interactive tool to investigate this exploration.

Work with a partner. A six-sided die is rolled. Draw a Venn diagram that relates the two events. Then decide whether the events are disjoint or overlapping.

a. Event A: The result is an even number.
Event B: The result is a prime number.

b. Event A: The result is 2 or 4.
Event B: The result is an odd number.

2 EXPLORATION: Finding the Probability that Two Events Occur

Work with a partner. A six-sided die is rolled. For each pair of events, find (a) $P(A)$, (b) $P(B)$, (c) $P(A \text{ and } B)$, and (d) $P(A \text{ or } B)$.

a. Event A: The result is an even number.
Event B: The result is a prime number.

b. Event A: The result is a 2 or 4.
Event B: The result is an odd number.

3 EXPLORATION: Discovering Probability Formulas

Go to *BigIdeasMath.com* for an interactive tool to investigate this exploration.

Work with a partner.

a. In general, if event A and event B are disjoint, then what is the probability that event A or event B will occur? Use a Venn diagram to justify your conclusion.

b. In general, if event A and event B are overlapping, then what is the probability that event A or event B will occur? Use a Venn diagram to justify your conclusion.

c. Conduct an experiment using a six-sided die. Roll the die 50 times and record the results. Then use the results to find the probabilities described in Exploration 2. How closely do your experimental probabilities compare to the theoretical probabilities you found in Exploration 2?

Communicate Your Answer

4. How can you find probabilities of disjoint and overlapping events?

5. Give examples of disjoint events and overlapping events that do not involve dice.

Name ______________________________ Date __________

10.4 Notetaking with Vocabulary

For use after Lesson 10.4

In your own words, write the meaning of each vocabulary term.

compound event

overlapping events

disjoint or mutually exclusive events

Core Concepts

Probability of Compound Events

If A and B are any two events, then the probability of A or B is

$$P(A \text{ or } B) = P(A) + P(B) - P(A \text{ and } B).$$

If A and B are disjoint events, then the probability of A or B is

$$P(A \text{ or } B) = P(A) + P(B).$$

Notes:

Extra Practice

1. Events A and B are disjoint. $P(A) = \frac{2}{3}$ and $P(B) = \frac{1}{6}$. Find $P(A \text{ or } B)$.

2. $P(A) = 0.8$, $P(B) = 0.05$, and $P(A \text{ or } B) = 0.6$. Find $P(A \text{ and } B)$.

In Exercises 3–6, a vehicle is randomly chosen from a parking lot. The parking lot contains three red minivans, two blue minivans, three blue convertibles, one black pickup truck, three black motorcycles, one red motorcycle and two blue scooters. Find the probability of selecting the type of vehicle.

3. A red vehicle or a minivan

4. A scooter or a black vehicle

5. A black vehicle or a motorcycle

6. A four-wheeled vehicle or a blue vehicle

7. During a basketball game, the coach needs to select a player to make the free throw after a technical foul on the other team. There is a 68% chance that the coach will select you and a 26% chance that the coach will select your friend. What is the probability that you or your friend is selected to make the free throw?

8. Two six-sided dice are rolled. Find the probability of rolling the same number twice.

9. Out of 120 student parents, 90 of them can chaperone the Homecoming dance or the Prom. There are 40 parents who can chaperone the Homecoming dance and 65 parents who can chaperone the Prom. What is the probability that a randomly selected parent can chaperone both the Homecoming dance and the Prom?

10. A football team scores a touchdown first 75% of the time when they start with the ball. The team does not score first 51% of the time when their opponent starts with the ball. The team who gets the ball first is determined by a coin toss. What is the probability that the team scores a touchdown first?

Name________________________________ Date__________

10.5 Permutations and Combinations

For use with Exploration 10.5

Essential Question How can a tree diagram help you visualize the number of ways in which two or more events can occur?

1 EXPLORATION: Reading a Tree Diagram

Work with a partner. Two coins are flipped and the spinner is spun. The tree diagram shows the possible outcomes.

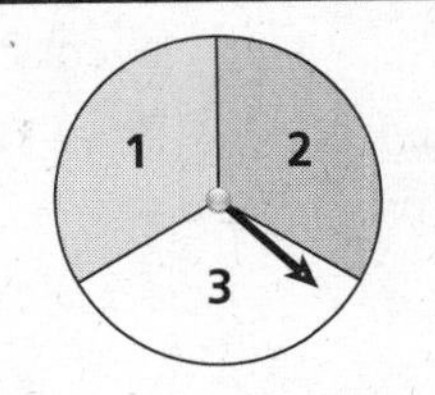

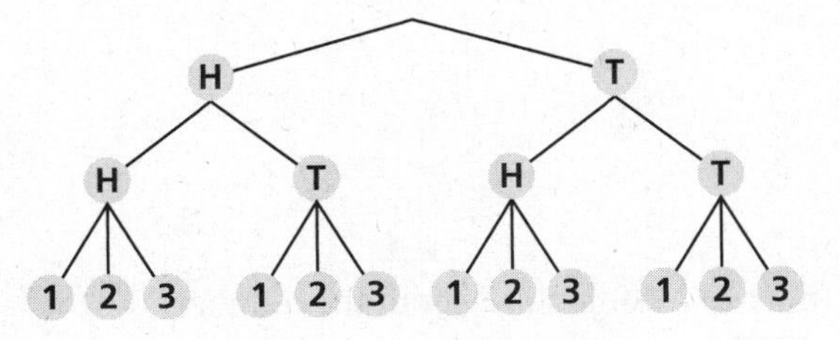

Coin is flipped.

Coin is flipped.

Spinner is spun.

a. How many outcomes are possible?

b. List the possible outcomes.

2 EXPLORATION: Reading a Tree Diagram

Work with a partner. Consider the tree diagram below.

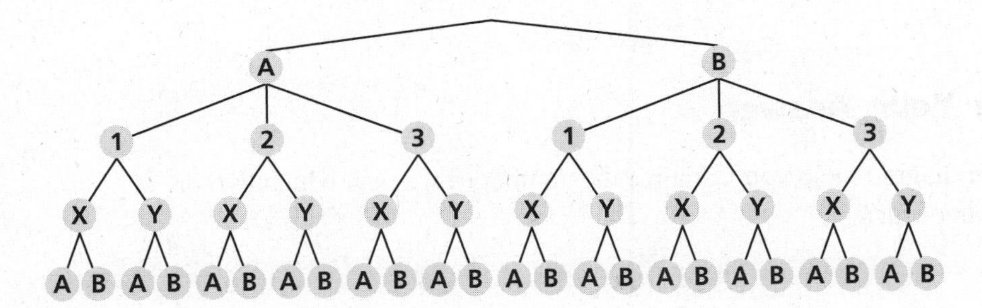

a. How many events are shown?

b. What outcomes are possible for each event?

c. How many outcomes are possible?

d. List the possible outcomes.

Name ______________________________ Date __________

10.5 Permutations and Combinations (continued)

3 EXPLORATION: Writing a Conjecture

Work with a partner.

a. Consider the following general problem: Event 1 can occur in m ways and event 2 can occur in n ways. Write a conjecture about the number of ways the two events can occur. Explain your reasoning.

b. Use the conjecture you wrote in part (a) to write a conjecture about the number of ways *more than* two events can occur. Explain your reasoning.

c. Use the results of Explorations 1(a) and 2(c) to verify your conjectures.

Communicate Your Answer

4. How can a tree diagram help you visualize the number of ways in which two or more events can occur?

5. In Exploration 1, the spinner is spun a second time. How many outcomes are possible?

Name____ Date__________

10.5 Notetaking with Vocabulary
For use after Lesson 10.5

In your own words, write the meaning of each vocabulary term.

permutation

n factorial

combination

Binomial Theorem

Core Concepts

Permutations

Formulas	**Examples**
The number of permutations of n objects is given by $_nP_n = n!.$	The number of permutations of 4 objects is $_4P_4 = 4! = 4 \bullet 3 \bullet 2 \bullet 1 = 24.$
The number of permutations of n objects taken r at a time, where $r \leq n$, is given by $_nP_r = \dfrac{n!}{(n-r)!}.$	The number of permutations of 4 objects taken 2 at a time is $_4P_2 = \dfrac{4!}{(4-2)!} = \dfrac{4 \bullet 3 \bullet \cancel{2!}}{\cancel{2!}} = 12.$

Notes:

Name ______________________________ Date __________

10.5 Notetaking with Vocabulary (continued)

Combinations

Formula The number of combinations of n objects taken r at a time, where $r \leq n$, is given by

$$_nC_r = \frac{n!}{(n-r)! \bullet r!}.$$

Example The number of combinations of 4 objects taken 2 at a time is

$$_4C_2 = \frac{4!}{(4-2)! \bullet 2!} = \frac{4 \bullet 3 \bullet \cancel{2!}}{\cancel{2!} \bullet (2 \bullet 1)} = 6.$$

Notes:

The Binomial Theorem

For any positive integer n, the binomial expansion of $(a + b)^n$ is

$$(a + b)^n = {}_nC_0 a^n b^0 + {}_nC_1 a^{n-1} b^1 + {}_nC_2 a^{n-2} b^2 + \cdots + {}_nC_n a^0 b^n.$$

Notice that each term in the expansion of $(a + b)^n$ has the form ${}_nC_r a^{n-r} b^r$, where r is an integer from 0 to n.

Notes:

Name______________________________ Date__________

Extra Practice

In Exercises 1 and 2, find the number of ways you can arrange (a) all of the numbers and (b) 3 of the numbers in the given amount.

1. \$2,564,783

2. \$4,128,675,309

3. Your rock band has nine songs recorded but you only want to put five of them on your demo CD to hand out to local radio stations. How many possible ways could the five songs be ordered on your demo CD?

4. A witness at the scene of a hit-and-run accident saw that the car that caused the accident had a license plate with only the letters I, R, L, T, O, and A. Find the probability that the license plate starts with a T and ends with an R.

5. How many possible combinations of three colors can be chosen from the seven colors of the rainbow?

6. Use the Binomial Theorem to write the binomial expansion of $(2x^4 + y^3)^3$.

Name ____ Date __________

10.6 Binomial Distributions
For use with Exploration 10.6

Essential Question How can you determine the frequency of each outcome of an event?

1 EXPLORATION: Analyzing Histograms

Go to *BigIdeasMath.com* for an interactive tool to investigate this exploration.

Work with a partner. The histograms show the results when n coins are flipped.

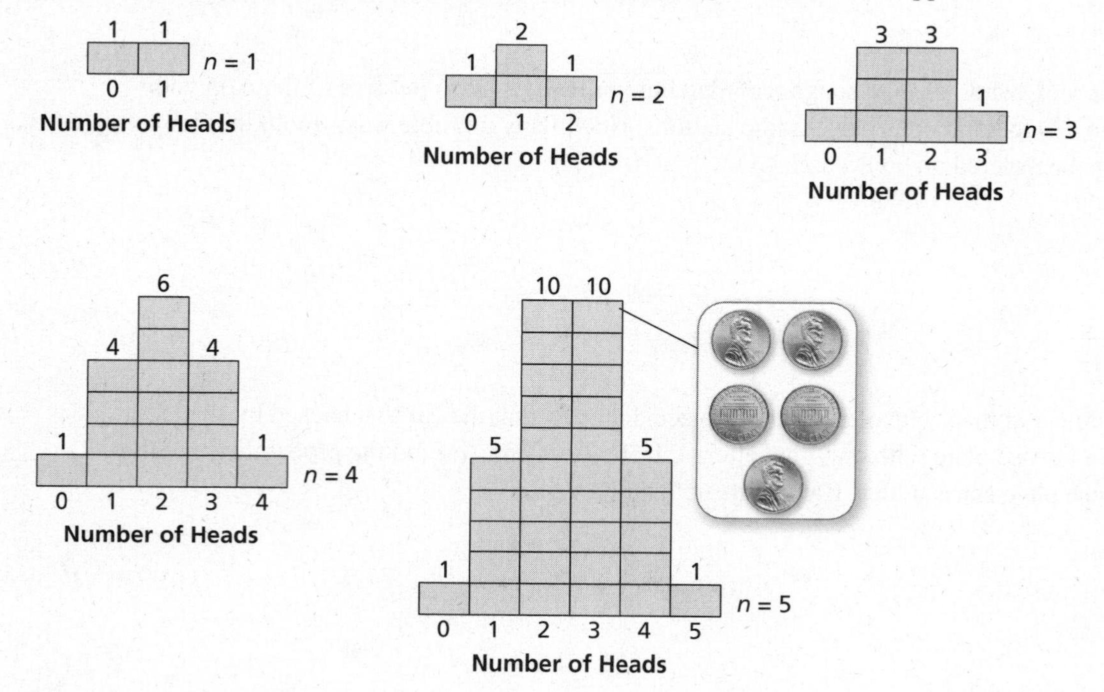

a. In how many ways can 3 heads occur when 5 coins are flipped?

b. Draw a histogram that shows the numbers of heads that can occur when 6 coins are flipped.

c. In how many ways can 3 heads occur when 6 coins are flipped?

Name____ Date__________

2 EXPLORATION: Determining the Number of Occurrences

Work with a partner.

a. Complete the table showing the numbers of ways in which 2 heads can occur when n coins are flipped.

n	3	4	5	6	7
Occurrences of 2 heads					

b. Determine the pattern shown in the table. Use your result to find the number of ways in which 2 heads can occur when 8 coins are flipped.

Communicate Your Answer

3. How can you determine the frequency of each outcome of an event?

4. How can you use a histogram to find the probability of an event?

Name ______________________________ Date __________

10.6 Notetaking with Vocabulary
For use after Lesson 10.6

In your own words, write the meaning of each vocabulary term.

random variable

probability distribution

binomial distribution

binomial experiment

Core Concepts

Probability Distributions

A **probability distribution** is a function that gives the probability of each possible value of a random variable. The sum of all the probabilities in a probability distribution must equal 1.

Probability Distribution for Rolling a Six-Sided Die						
x	1	2	3	4	5	6
$P(x)$	$\frac{1}{6}$	$\frac{1}{6}$	$\frac{1}{6}$	$\frac{1}{6}$	$\frac{1}{6}$	$\frac{1}{6}$

Notes:

10.6 Notetaking with Vocabulary (continued)

Binomial Experiments

A **binomial experiment** meets the following conditions.

- There are n independent trials.
- Each trial has only two possible outcomes: success and failure.
- The probability of success is the same for each trial. This probability is denoted by p. The probability of failure is $1 - p$.

For a binomial experiment, the probability of exactly k successes in n trials is

$$P(k \text{ successes}) = {}_nC_k p^k (1 - p)^{n-k}.$$

Notes:

Extra Practice

1. Make a table and draw a histogram showing the probability distribution for the random variable P if P = the product when two six-sided dice are rolled.

Name ______________________________ Date __________

10.6 Notetaking with Vocabulary (continued)

2. Use the probability distribution to determine (a) the number that is most likely to be spun on a spinner, and (b) the probability of spinning a perfect square.

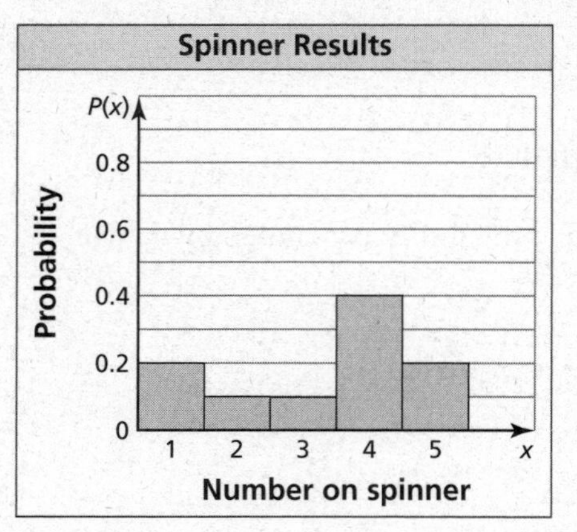

3. Calculate the probability of flipping a coin twenty times and getting nineteen heads.

4. According to a survey, 78% of women in a city watch professional football. You ask four randomly chosen women from the city whether they watch professional football.

a. Draw a histogram of the binomial distribution for your survey.

b. What is the most likely outcome of your survey?

c. What is the probability that at most one woman watches professional football?

Name____ Date__________

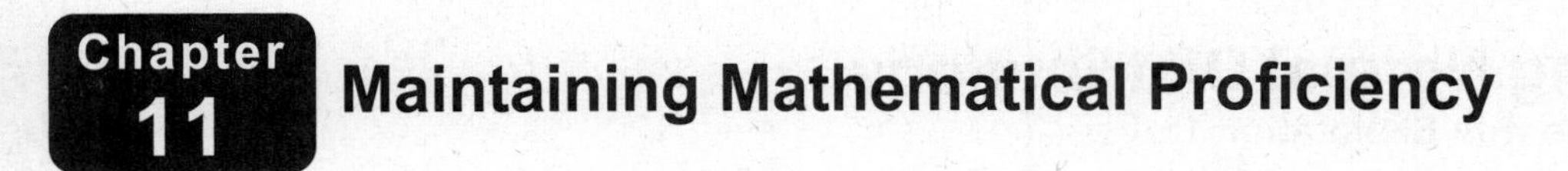

Maintaining Mathematical Proficiency

Find the mean, median, and mode of the data set. Then determine which measure of center best represents the data. Explain.

1. 26, 24, 55, 21, 32, 26

2. 63, 66, 61, 70, 69, 67, 63, 65

3. 40, 37, 21, 43, 37, 41, 43, 25, 37

Find and interpret the standard deviation of the data set.

4. 18, 11, 15, 20, 16

5. 78, 71, 68, 75, 46, 66

Name ______________________________ Date __________

11.1 Using Normal Distributions

For use with Exploration 11.1

Essential Question In a normal distribution, about what percent of the data lies within one, two, and three standard deviations of the mean?

Recall that the standard deviation σ of a numerical data set is given by

$$\sigma = \sqrt{\frac{(x_1 - \mu)^2 + (x_2 - \mu)^2 + \cdots + (x_n - \mu)^2}{n}}$$

where n is the number of values in the data set and μ is the mean of the data set.

EXPLORATION: Analyzing a Normal Distribution

Work with a partner. In many naturally occurring data sets, the histogram of the data is bell-shaped. In statistics, such data sets are said to have a *normal distribution*. For the normal distribution shown below, estimate the percent of the data that lies within one, two, and three standard deviations of the mean. Each square on the grid represents 1%.

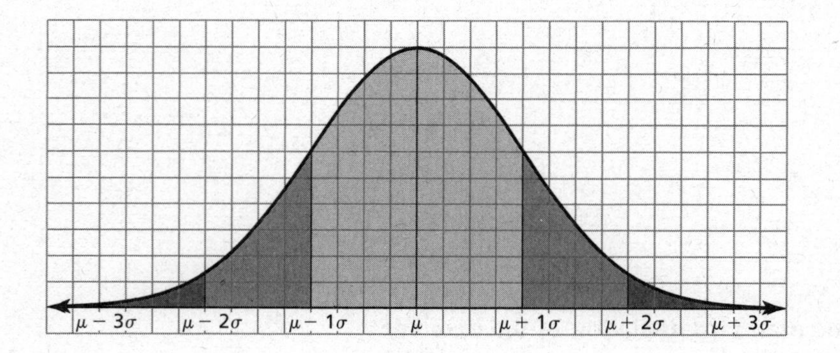

Name____ Date__________

11.1 Using Normal Distributions (continued)

2 EXPLORATION: Analyzing a Data Set

Work with a partner. A famous data set was collected in Scotland in the mid-1800s. It contains the chest sizes (in inches) of 5738 men in the Scottish Militia. Do the data fit a normal distribution? Explain.

Chest size	Number of men
33	3
34	18
35	81
36	185
37	420
38	749
39	1073
40	1079
41	934
42	658
43	370
44	92
45	50
46	21
47	4
48	1

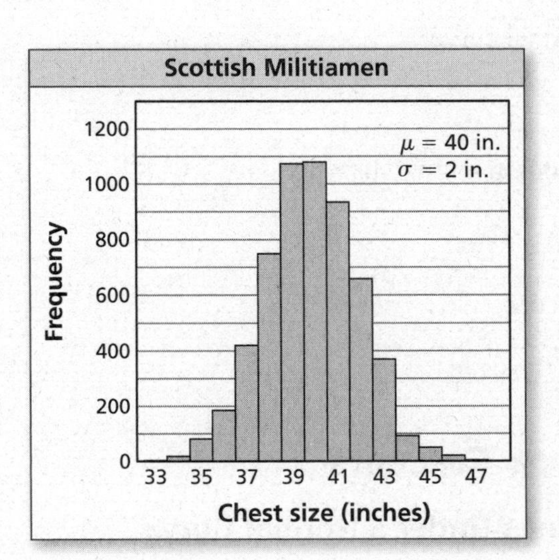

Communicate Your Answer

3. In a normal distribution, about what percent of the data lies within one, two, and three standard deviations of the mean?

4. Use the Internet or some other reference to find another data set that is normally distributed. Display your data in a histogram.

Name ______________________________ Date __________

11.1 Notetaking with Vocabulary

For use after Lesson 11.1

In your own words, write the meaning of each vocabulary term.

normal distribution

normal curve

standard normal distribution

z-score

Core Concepts

Areas Under a Normal Curve

A normal distribution with mean μ and standard deviation σ has these properties.

- The total area under the related normal curve is 1.
- About 68% of the area lies within 1 standard deviation of the mean.
- About 95% of the area lies within 2 standard deviations of the mean.
- About 99.7% of the area lies within 3 standard deviations of the mean.

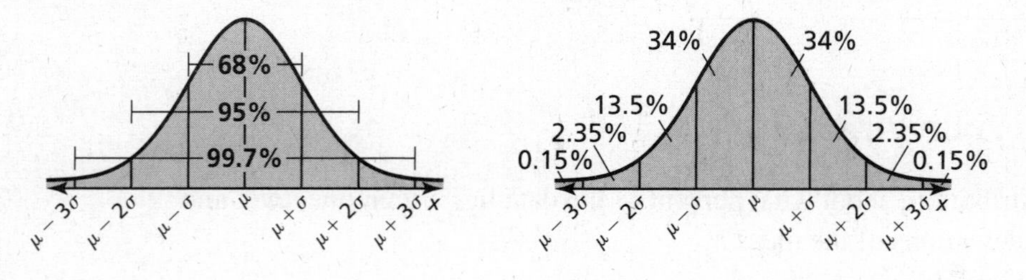

Notes:

Name____ Date __________

Extra Practice

In Exercises 1–6, a normal distribution has mean μ and standard deviation σ. Find the indicated probability for a randomly selected *x*-value from the distribution.

1. $P(x \leq \mu - 2\sigma)$

2. $P(x \geq \mu - 3\sigma)$

3. $P(x \leq \mu + 2\sigma)$

4. $P(x \geq \mu + 3\sigma)$

5. $P(\mu - \sigma \leq x \leq \mu + 3\sigma)$

6. $P(\mu - 2\sigma \leq x \leq \mu + \sigma)$

7. The scores for a math course test are normally distributed with a mean of 61 and a standard deviation of 11. The test scores range from 0 to 100.

a. About what percent of the students taking the test have scores between 72 and 83?

b. About what percent of the students taking the test have scores less than 50?

11.1 Notetaking with Vocabulary (continued)

8. The temperatures of a city are normally distributed over the course of a year. The mean temperature is 55.2°F and the standard deviation is 6.3°F. A day is randomly chosen.

a. What is the probability that the chosen day is 45°F or cooler?

b. What is the probability that the chosen day is cooler than 32.5°F?

c. What is the probability that the chosen day is between 32.5°F and 45°F?

d. What is the probability that the chosen day is 60°F or warmer?

In Exercises 9 and 10, determine whether the histogram has a normal distribution.

9. **10.**

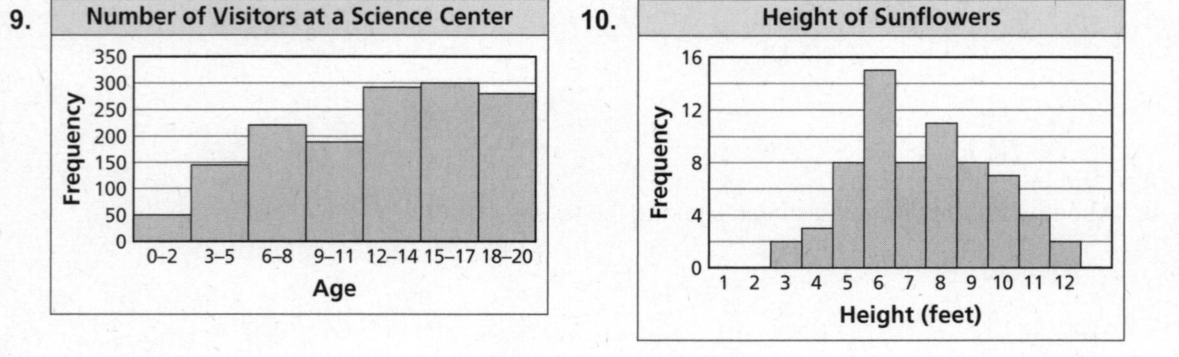

Name____ Date__________

11.2 Populations, Samples, and Hypotheses

For use with Exploration 11.2

Essential Question How can you test theoretical probability using sample data?

1 EXPLORATION: Using Sample Data

Go to *BigIdeasMath.com* for an interactive tool to investigate this exploration.

Work with a partner.

a. When two six-sided dice are rolled, what is the theoretical probability that you roll the same number on both dice?

b. Conduct an experiment to check your answer in part (a). What sample size did you use? Explain your reasoning.

c. Use the dice rolling simulator at *BigIdeasMath.com* to complete the table. Do your experimental data check the theoretical probability you found in part (a)? Explain. What happens as you increase the sample size?

Number of Rolls	Number of Times Same Number Appears	Experimental Probability
100		
500		
1000		
5000		
10,000		

2 EXPLORATION: Using Sample Data

Go to *BigIdeasMath.com* for an interactive tool to investigate this exploration.

Work with a partner.

a. When three six-sided dice are rolled, what is the theoretical probability that you roll the same number on all three dice?

Name ______________________________ Date __________

11.2 Populations, Samples, and Hypotheses (continued)

2 EXPLORATION: Using Sample Data (continued)

b. Compare the theoretical probability you found in part (a) with the theoretical probability you found in Exploration 1(a).

c. Conduct an experiment to check your answer in part (a). How does adding a die affect the sample size that you use? Explain your reasoning.

d. Use the dice rolling simulator at *BigIdeasMath.com* to check your answer to part (a). What happens as you increase the sample size?

Communicate Your Answer

3. How can you test theoretical probability using sample data?

4. Conduct an experiment to determine the probability of rolling a sum of 7 when two six-sided dice are rolled. Then find the theoretical probability and compare your answers.

Name____ Date__________

11.2 Notetaking with Vocabulary
For use after Lesson 11.2

In your own words, write the meaning of each vocabulary term.

population

sample

parameter

statistic

hypothesis

Notes:

Name ______________________________ Date __________

11.2 Notetaking with Vocabulary (continued)

Extra Practice

In Exercises 1–3, identify the population and sample. Describe the sample.

1. In a city, a survey of 3257 adults ages 18 and over found that 2605 of them own a tablet.

2. To find out the consumers' response towards a new flavor of sports drink, a company surveys 1000 athletes who drink sports drinks and finds that 726 of them like the new flavor.

3. In a school district, a survey of 1500 high school students found that 824 of them have a part time job in the summer.

In Exercises 4–7, determine whether the numerical value is a parameter or a statistic. Explain your reasoning.

4. Eighty-two percent of the residents in one neighborhood in a town voted to approve building a bike lane through town.

Name____ Date__________

5. In a science class, 25% of the students wear glasses.

6. In a recent year, the median household income in the United States was about $52,000.

7. A survey of some visitors to a museum found that 84% thought the new planetarium was very exciting.

8. You spin the spinner five times and every time the spinner lands on blue. You suspect the spinner favors blue. The maker of the spinner claims that the spinner does not favor any color. You simulate spinning the spinner 50 times by repeatedly drawing 200 random samples of size 50. The histogram shows the results. Use the histogram to determine what you should conclude when you spin the actual spinner 50 times and the spinner lands on blue (a) 12 times and (b) 19 times.

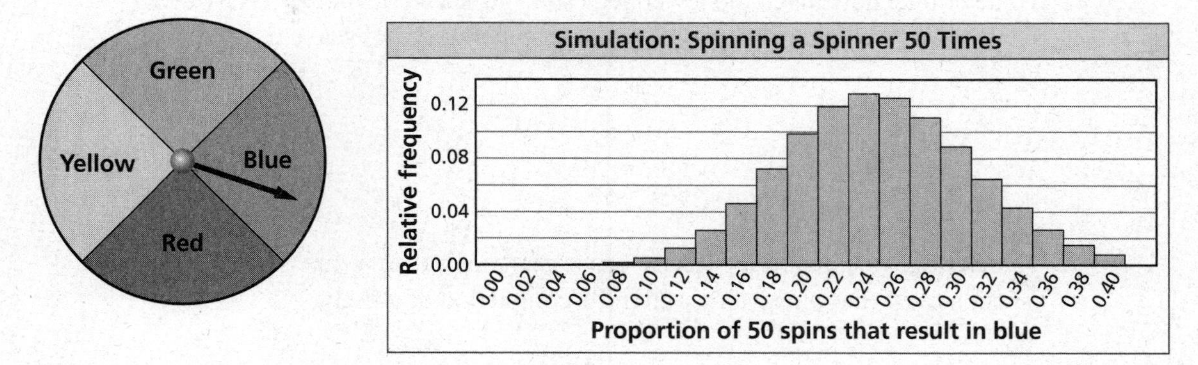

Name ______________________________ Date __________

11.3 Collecting Data

For use with Exploration 11.3

Essential Question What are some considerations when undertaking a statistical study?

EXPLORATION: Analyzing Sampling Techniques

Work with a partner. Determine whether each sample is representative of the population. Explain your reasoning.

a. To determine the number of hours people exercise during a week, researchers use random-digit dialing and call 1500 people.

b. To determine how many text messages high school students send in a week, researchers post a survey on a website and receive 750 responses.

c. To determine how much money college students spend on clothes each semester, a researcher surveys 450 college students as they leave the university library.

d. To determine the quality of service customers receive, an airline sends an e-mail survey to each customer after the completion of a flight.

EXPLORATION: Analyzing Survey Questions

Work with a partner. Determine whether each survey question is biased. Explain your reasoning. If so, suggest an unbiased rewording of the question.

a. Does eating nutritious, whole-grain foods improve your health?

2 EXPLORATION: Analyzing Survey Questions (continued)

b. Do you ever attempt the dangerous activity of texting while driving?

c. How many hours do you sleep each night?

d. How can the mayor of your city improve his or her public image?

3 EXPLORATION: Analyzing Survey Randomness and Truthfulness

Work with a partner. Discuss each potential problem in obtaining a random survey of a population. Include suggestions for overcoming the problem.

a. The people selected might not be a random sample of the population.

b. The people selected might not be willing to participate in the survey.

c. The people selected might not be truthful when answering the question.

d. The peopled selected might not understand the survey question.

Communicate Your Answer

4. What are some considerations when undertaking a statistical study?

5. Find a real-life example of a biased survey question. Then suggest an unbiased rewording of the question.

Name ______________________________ Date __________

11.3 Notetaking with Vocabulary

For use after Lesson 11.3

In your own words, write the meaning of each vocabulary term.

random sample

self-selected sample

systematic sample

stratified sample

cluster sample

convenience sample

bias

unbiased sample

biased sample

experiment

observational study

survey

11.3 Notetaking with Vocabulary (continued)

simulation

biased question

Core Concepts

Types of Samples

For a **self-selected sample**, members of a population can volunteer to be in the sample.

For a **systematic sample**, a rule is used to select members of a population. For instance, selecting every other person.

For a **stratified sample**, a population is divided into smaller groups that share a similar characteristic. A sample is then randomly selected from each group.

For a **cluster sample**, a population is divided into groups, called *clusters*. All of the members in one or more of the clusters are selected.

For a **convenience sample**, only members of a population who are easy to reach are selected.

11.3 Notetaking with Vocabulary (continued)

Methods of Collecting Data

An **experiment** imposes a treatment on individuals in order to collect data on their response to the treatment. The treatment may be a medical treatment, or it can be any action that might affect a variable in the experiment, such as adding methanol to gasoline and then measuring its effect on fuel efficiency.

An **observational study** observes individuals and measures variables without controlling the individuals or their environment. This type of study is used when it is difficult to control or isolate the variable being studied, or when it may be unethical to subject people to a certain treatment or to withhold it from them.

A **survey** is an investigation of one or more characteristics of a population. In a survey, every member of a sample is asked one or more questions.

A **simulation** uses a model to reproduce the conditions of a situation or process so that the simulated outcomes closely match the real-world outcomes. Simulations allow you to study situations that are impractical or dangerous to create in real life.

Notes:

Extra Practice

In Exercises 1–3, identify the type of sample described.

1. A restaurant owner wants to know whether the customers are satisfied with the service. Every fifth customer who exits the restaurant is surveyed.

2. An electronic manufacturer wants to know the customers' responses towards a newly released media player. Emails are sent to customers who recently purchased the device to participate in an online survey at their convenience.

3. A survey is conducted in a state to find out how many households own more than one vehicle. Households are divided into north, east, south, and west regions of the state, and a sample is randomly surveyed from each region.

In Exercises 4 and 5, identify the type of sample and explain why the sample is biased.

4. A manager of a company wants to determine whether the employees are satisfied with the lounge room. The manager surveys the employees who are in the lounge room during lunch break.

5. A news station asks its viewers to participate in an online poll about the presidential candidates.

In Exercises 6 and 7, identify the method of data collection the situation describes.

6. A researcher records whether shoppers at a grocery store buy magazines at the checkout aisles while waiting in line to check out.

7. A meteorologist uses a computer model to track the trajectory of a hurricane.

Name ______________________________ Date __________

11.4 Experimental Design

For use with Exploration 11.4

Essential Question How can you use an experiment to test a conjecture?

1 EXPLORATION: Using an Experiment

Work with a partner. Standard white playing dice are manufactured with black dots that are indentations, as shown. So, the side with six indentations is the lightest side and the side with one indentation is the heaviest side.

You make a conjecture that when you roll a standard playing die, the number 6 will come up more often because it is the lightest side, and the number 1 will come up least often because it is the heaviest side. To test your conjecture, roll a standard playing die 25 times. Record the results in the table. Does the experiment confirm your conjecture? Explain your reasoning.

Number						
Rolls						

Name____ Date__________

11.4 Experimental Design (continued)

2 EXPLORATION: Analyzing an Experiment

Work with a partner. To overcome the imbalance of standard playing dice, one of the authors of this book invented and patented 12-sided dice, on which each number from 1 through 6 appears twice (on opposing sides). See *BigIdeasMath.com*.

As part of the patent process, a standard playing die was rolled 27,090 times. The results are shown below.

Number	1	2	3	4	5	6
Rolls	4293	4524	4492	4397	4623	4761

What can you conclude from the results of this experiment? Explain your reasoning.

Communicate Your Answer

3. How can you use an experiment to test a conjecture?

4. Make a conjecture about the outcomes of rolling the 12-sided die in Exploration 2. Then use the Internet to find a 12-sided die rolling simulator. Use the simulator to complete the table. How many times did you simulate rolling the die? Explain your reasoning.

Number						
Rolls						

Name ______________________________ Date ________

11.4 Notetaking with Vocabulary
For use after Lesson 11.4

In your own words, write the meaning of each vocabulary term.

controlled experiment

control group

treatment group

randomization

randomized comparative experiment

placebo

replication

Core Concepts

Comparative Studies and Causality

- A rigorous randomized comparative experiment, by eliminating sources of variation other than the controlled variable, can make valid cause-and-effect conclusions possible.
- An observational study can identify *correlation* between variables, but not *causality*. Variables, other than what is being measured, may be affecting the results.

Notes:

Name____ Date__________

Extra Practice

In Exercises 1 and 2, determine whether the study is a randomized comparative experiment. If it is, describe the treatment, the treatment group, and the control group. If it is not, explain why not and discuss whether the conclusions drawn from the study are valid.

1.

Baby DVDs

Baby DVDs Improves Language Ability

To test whether baby DVDs that highlight words and introduce music and art can improve language ability, parents with babies 0–24 months were given the choice of whether to let their babies watch the DVDs. Fifty babies who watched the DVDs were observed for a year as well as 50 other babies who did not watch the DVDs. At the end of the year, babies who watched the DVDs scored higher in a language development test.

2.

Type 1 Diabetes

New Drug Improves Blood Glucose Control

In a clinical trial, 100 Type 1 diabetic patients volunteered to take a new drug. Fifty percent of the patients received the drug and the other fifty percent received a placebo. After one year, the patients who received the drug had better blood glucose control while the placebo group experienced no significant change.

11.4 Notetaking with Vocabulary (continued)

In Exercises 3 and 4, explain whether the research topic is best investigated through an experiment or an observational study. Then describe the design of the experiment or observational study.

3. A criminologist wants to know whether social factors are the cause of the criminal behavior.

4. A pharmaceutical company wants to know whether the new medication on heart disease has a side effect on individuals.

5. A company wants to test the effectiveness of a new moisturizing cream designed to help improve skin complexion. Identify a potential problem, if any, with each experimental design. Then describe how you can improve it.

a. The company randomly selects ten individuals. Five subjects are given the new moisturizing cream and the other five are given a placebo. After eight weeks, each subject is evaluated and it is determined that the five subjects who have been using the cream have improved skin complexion.

b. The company randomly selects a large group of individuals. Half of the individuals are given the new moisturizing cream and the other half of the individuals may use their own existing moisturizers or none at all. After eight weeks, each subject is evaluated and it is determined that a significant large number of subjects who received the moisturizing cream have improved skin complexion.

Name ____________________ Date __________

11.5 Making Inferences from Sample Surveys

For use with Exploration 11.5

Essential Question How can you use a sample survey to infer a conclusion about a population?

1 EXPLORATION: Making an Inference from a Sample

Go to *BigIdeasMath.com* for an interactive tool to investigate this exploration.

Work with a partner. You conduct a study to determine what percent of the high school students in your city would prefer an upgraded model of their current cell phone. Based on your intuition and talking with a few acquaintances, you think that 50% of high school students would prefer an upgrade. You survey 50 randomly chosen high school students and find that 20 of them prefer an upgraded model.

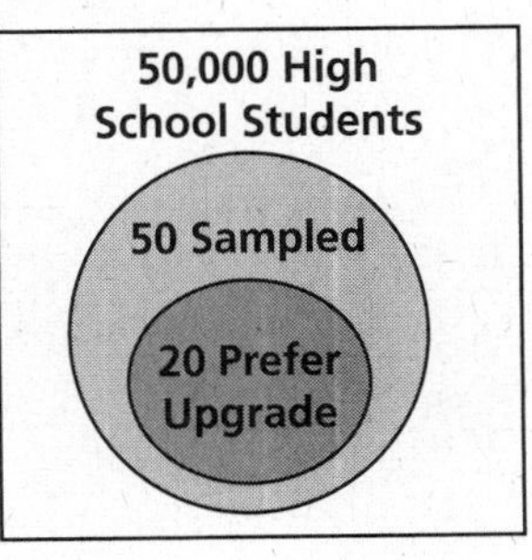

a. Based on your sample survey, what percent of the high school students in your city would prefer an upgraded model? Explain your reasoning.

b. In spite of your sample survey, is it still possible that 50% of the high school students in your city prefer an upgraded model? Explain your reasoning.

c. To investigate the likelihood that you could have selected a sample of 50 from a population in which 50% of the population does prefer an upgraded model, you create a binomial distribution as shown below. From the distribution, estimate the probability that exactly 20 students surveyed prefer an upgraded model. Is this event likely to occur? Explain your reasoning.

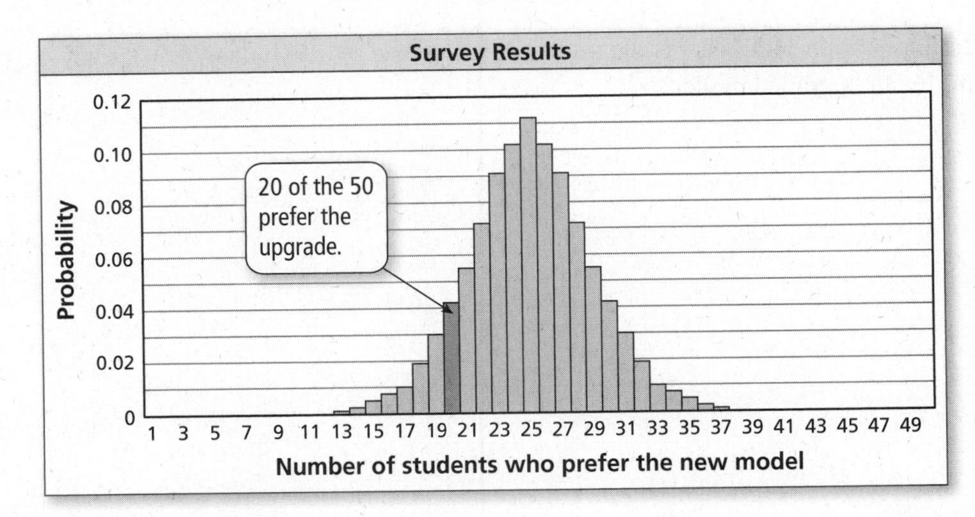

Name ____ Date __________

1 EXPLORATION: Making an Inference from a Sample (continued)

d. When making inferences from sample surveys, the sample must be random. In the situation described on the previous page, describe how you could design and conduct a survey using a random sample of 50 high school students who live in a large city.

Communicate Your Answer

2. How can you use a sample survey to infer a conclusion about a population?

3. In Exploration 1(c), what is the probability that exactly 25 students you survey prefer an upgraded model?

Name____ Date__________

11.5 Notetaking with Vocabulary

For use after Lesson 11.5

In your own words, write the meaning of each vocabulary term.

descriptive statistics

inferential statistics

margin of error

Core Concepts

Margin of Error Formula

When a random sample of size n is taken from a large population, the margin of error is approximated by

$$\text{Margin of error} = \pm\frac{1}{\sqrt{n}}.$$

This means that if the percent of the sample responding a certain way is p (expressed as a decimal), then the percent of the population who would respond the same way is likely to be between $p - \frac{1}{\sqrt{n}}$ and $p + \frac{1}{\sqrt{n}}$.

Notes:

Name ______________________ Date ________

11.5 Notetaking with Vocabulary (continued)

Extra Practice

1. The numbers of minutes spent each day on a social networking website by a random sample of people between the ages of 18 and 64 are shown in the table. Estimate the population mean μ.

Number of Minutes				
175	15	190	180	45
100	210	240	190	60
102	165	253	192	102
12	180	189	193	230
300	185	190	395	186
183	200	165	195	409

2. Use the data in Exercise 1 to answer each question.

   a. Estimate the population proportion ρ of social network users between the ages of 18 and 64 who spend more than 120 minutes each day on a social networking website.

   b. Estimate the population proportion ρ of social network users between the ages of 18 and 64 who spend fewer than 60 minutes each day on a social networking website.

Name______________________________ Date__________

11.5 Notetaking with Vocabulary (continued)

3. Two candidates, A and B, are running for the student council president position. The table shows the results from four surveys of randomly selected students in the school. The students are asked whether they will vote for candidate A. The results are shown in the table.

Sample Size	Number of Votes for Candidate A	Percent of Votes for Candidate A
10	6	60%
20	11	55%
50	20	40%
150	64	42.7%

a. Based on the results of the first two surveys, do you think Candidate A will win the election? Explain.

b. Based on the results in the table, do you think Candidate A will win the election? Explain.

4. A national polling company claims that 39% of Americans rate the overall quality of the environment in the nation as "good." You survey a random sample of 50 people. What can you conclude about the accuracy of the claim that the population proportion is 0.39 when 19 Americans say the quality of the environment is good?

5. In a survey of 2680 people in the U.S., 60% said that their diet is somewhat healthy.

a. What is the margin of error for the survey?

b. Give an interval that is likely to contain the exact percent of all people in the U.S. who think their diet is somewhat healthy.

Name ______________________________ Date ________

11.6 Making Inferences from Experiments

For use with Exploration 11.6

Essential Question How can you test a hypothesis about an experiment?

1 EXPLORATION: Resampling Data

Go to *BigIdeasMath.com* for an interactive tool to investigate this exploration.

Work with a partner. A randomized comparative experiment tests whether water with dissolved calcium affects the yields of yellow squash plants. The table shows the results.

Yield (kilograms)	
Control Group	**Treatment Group**
1.0	1.1
1.2	1.3
1.5	1.4
0.9	1.2
1.1	1.0
1.4	1.7
0.8	1.8
0.9	1.1
1.3	1.1
1.6	1.8

a. Find the mean yield of the control group and the mean yield of the treatment group. Then find the difference of the two means. Record the results.

b. Write each yield measurement from the table on an equal-sized piece of paper. Place the pieces of paper in a bag, shake, and randomly choose 10 pieces of paper. Call this the "control" group, and call the 10 pieces in the bag the "treatment" group. Then repeat part (a) and return the pieces to the bag. Perform this resampling experiment five times.

c. How does the difference in the means of the control and treatment groups compare with the differences resulting from chance?

2 EXPLORATION: Evaluating Results

Work as a class. To conclude that the treatment is responsible for the difference in yield, you need strong evidence to reject the hypothesis:

Water dissolved in calcium has no effect on the yields of yellow squash plants.

To evaluate this hypothesis, compare the experimental difference of means with the resampling differences.

a. Collect all the resampling differences of means found in Exploration 1(b) for the whole class and display these values in a histogram.

b. Draw a vertical line on your class histogram to represent the experimental difference of means found in Exploration 1(a).

c. Where on the histogram should the experimental difference of means lie to give evidence for rejecting the hypothesis?

d. Is your class able to reject the hypothesis? Explain your reasoning.

Communicate Your Answer

3. How can you test a hypothesis about an experiment?

4. The randomized comparative experiment described in Exploration 1 is replicated and the results are shown in the table. Repeat Explorations 1 and 2 using this data set. Explain any differences in your answers.

	Yield (kilograms)									
Control Group	0.9	0.9	1.4	0.6	1.0	1.1	0.7	0.6	1.2	1.3
Treatment Group	1.0	1.2	1.2	1.3	1.0	1.8	1.7	1.2	1.0	1.9

Name ______________________________ Date __________

11.6 Notetaking with Vocabulary
For use after Lesson 11.6

In your own words, write the meaning of each vocabulary term.

randomized comparative experiment

control group

treatment group

mean

dot plot

outlier

simulation

hypothesis

Notes:

Extra Practice

1. A randomized comparative experiment tests whether students who are given weekly quizzes do better on the comprehensive final exam. The control group has 10 students and the treatment group, which receives weekly quizzes, has 10 students. The table shows the results.

	Final Exam Scores (out of 100 points)									
Control Group	82	55	76	92	76	76	82	58	69	79
Treatment Group	92	90	88	73	88	63	94	81	81	77

a. Find the mean score of the control group.

b. Find the mean score of the treatment group.

c. Find the experimental difference of the means.

d. Display the data in a double dot plot.

e. What can you conclude?

Name ______________________________ Date __________

11.6 Notetaking with Vocabulary (continued)

2. Resample the data in Exercise 1 using a simulation. Use the means of the new control and treatment groups to calculate the difference of the means.

	Final Exam Scores (out of 100 points)									
New Control Group										
New Treatment Group										

3. To analyze the hypothesis below, use the histogram which shows the results from 200 resamplings of the data in Exercise 1.

Weekly Quizzes have no effect on final exam scores.

Compare the experimental difference in Exercise 1 with the resampling differences. What can you conclude about the hypothesis? Do weekly quizzes have an effect on final exam scores?

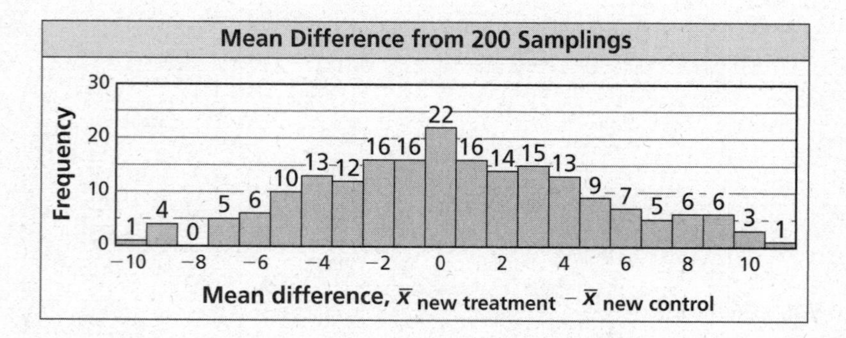